T0320655

MARKEDNESS

MARKEDNESS

Edited by
FRED R. ECKMAN
EDITH A. MORAVCSIK
and
JESSICA R. WIRTH
The University of Wisconsin–Milwaukee
Milwaukee, Wisconsin

PLENUM PRESS • NEW YORK AND LONDON

Library of Congress Cataloging in Publication Data

Linguistics Symposium of the University of Wisconsin–Milwaukee (12th: 1983)
  Markedness.

  "Proceedings of the Twelfth Annual Linguistics Symposium of the University of
Wisconsin–Milwaukee, held March 11–12, 1983, at the University of Wisconsin–
Milwaukee, Milwaukee, Wisconsin"—T.p. verso.
  Includes bibliographies and indexes.
  1. Markedness (Linguistics)—Congresses. I. Eckman, Fred R. II. Moravcsik, Edith
A. III. Wirth, Jessica. IV. Title.
P299.M35L54   1983   415   86-15096
ISBN 0-306-42372-3

Proceedings of the Twelfth Annual Linguistics Symposium of the University
of Wisconsin–Milwaukee, held March 11–12, 1983, at the University of
Wisconsin–Milwaukee, Milwaukee, Wisconsin

© 1986 Plenum Press, New York
A Division of Plenum Publishing Corporation
233 Spring Street, New York, N.Y. 10013

PREFACE

This volume presents the proceedings of the Twelfth Annual Linguistics Symposium of the University of Wisconsin-Milwaukee held March 11-12, 1983 on the campus of UWM.

It includes all papers that were given at the conference with the exception of Genevieve Escure and Glenn Gilbert's joint paper "Syntactic marking/unmarking phenomena in the creole continuum of Belize" which was not submitted for publication by the authors. Many of the papers appear in this volume in a revised form that is somewhat different from the oral version.

We would like to thank the various departments and other units at the University of Wisconsin-Milwaukee that sponsored the markedness symposium. These are: the Department of Linguistics, the English as a Second Language Intensive Program, the College of Letters and Science, the Division of Urban Outreach, the Center for Latin America and the Spanish Speaking Outreach Institute.

Finally, we wish to thank Lisa Carrara for doing a careful job on the preparation of the index, and J. L. Russell, for his patience and perseverance in typing a difficult manuscript.

F.R. Eckman,
E.A. Moravcsik and
J.R. Wirth

CONTENTS

MARKEDNESS - AN OVERVIEW

Edith Moravcsik and Jessica Wirth

University of Wisconsin-Milwaukee

1. Introduction

There is a relationship between the degree of our familiarity
with things and the number of subdistinctions we can perceive among
them. The better we know things the more we are able to detect secondary characteristics which assign them into sub-classes. This is
widely illustrable perhaps from all areas of human perception. For
example, Caucasians who do not often encounter Asians will perceive
all Orientals as looking alike. Or: people who have never consci-
ously observed bird songs will hear most birds as sounding the same.
Besides conceptual rapport, simple temporal distance also blurs
differences among things. Thus, events that are long past in one's
life may lose their special characteristics: all childhood friends
may later be recalled as more or less indistinct members of the general class. And, of course, spatial distance itself also obscures
subdistinctions: trees may seem all alike if viewed at great distance such as from a mountain top whereas, when seen at close range,
they turn out to represent many subtypes.

This correlation between familiarity and variability is apparent
not only in how people perceive things but also in how they construct
objects. Artifacts in everyday use - such as everyday clothing or
everyday foods - come in many more kinds than artifacts that have
special designations such as festive attire and holiday food. As a
trip to any greeting card shop would indicate, birthday cards or
Christmas cards come in many more kinds than graduation cards or St.
Patrick's Day cards.

Besides variability, there is another factor that also shows a

1

fairly consistent correlation with familiarity. This factor is
structural complexity. The relation is inverse in this case: the
more common an object is in our experience, the more simple it is
perceived or created to be. Caucasians often characterize Oriental
physiognomy as having additional characteristics - such as yellow
skin and almond-shaped eyes - that are added, as it were, to simple
Caucasian facial structure. And everyday clothing, everyday food,
and basic-purpose buildings tend to have a simpler structure than
festive clothing, holiday food, and special-purpose architecture.

There is thus a three-way correlation manifest in these examples
among familiarity, variability, and complexity. Closer familiarity
tends to be paired with simpler structure and greater variability;
less frequent occurrence in human experience goes with increased
structural complexity and diminished variability. People tend to
see familiar objects simpler in structure and more varied in kind
than less familiar objects; and they tend to make things that are in
common use to be simpler in structure and more variegated than those
in less common use.

This simple and intuitively not unreasonable three-way correlation is abundantly instantiated in the domain of language as well.
For a clear case, let us consider the expression of singular and
plural numbers. In many languages, singular has a morphologically
simpler expression and perhaps in no language is it morphologically
more complex than the plural. Corresponding to this simplicity of
form, frequency counts show singular forms to occur in greater numbers in texts than plural forms (cp. e.g. Greenberg 1966, 31f) -
they are thus more common, more familiar. And, thirdly, singular
forms exhibit in many languages a greater differentiation in terms
of gender or case than plural forms do. This is so in German, for
instance, where articles show a three-way gender differentiation in
the singular but not in the plural. In sum: the singular tends to
be simpler in form, more common in usage, and more elaborated in
terms of subtypes.

For another linguistic example, take voiced and voiceless stops.
Voiceless stops require less complex articulation when pronounced in
isolation than voiced ones where added vocal cord activity steps in.
They are also more widely distributed across languages in that there
are languages with only voiceless but no voiced stops but possibly
none than have voiced stops without voiceless ones.[1] And in languages like English at least, voiceless stops are more varied in
subtypes: they come both aspirated and unaspirated, as opposed to
the voiced ones which come in one variety only. In both the morphological and the phonetic example, a correlation is evident among
three logically independent factors: distribution, syntagmatic composition, and paradigmatic composition. In each case, of a pair of
opposing forms - singular-plural and voiceless-voiced - the one that
is crosslinguistically or language-internally more widely distributed

is also the one that is syntagmatically simpler and paradigmatically
more complex.

It is the recognition of these correlations in language structure
that constitutes the classic form of markedness theory. Following
the Prague School linguists, Greenberg (1966) assigns the designations "marked" and "unmarked" to opposing structural entities that
exhibit a consistently asymmetric relationship in terms of distribution and/or syntagmatic structure and/or paradigmatic complexity.
The one of the two entities that is consistently more widely distributed and/or simpler and/or more richly elaborated is called "unmarked"; its complement is the "marked" members of the opposition.
The singular number is thus unmarked as opposed to the plural which
is marked; and voiceless stops are unmarked as against voiced stops
which are marked.

Although probably all versions of markedness theory follow the
classic version in making claims about correlations within or across
the three domains of distribution, syntagmatic complexity, and paradigmatic complexity, they also differ widely in the more specific
content of their hypotheses. The papers presented at the '83 Milwaukee conference on markedness, the proceedings of which this volume
presents, amply illustrate the variety of ways in which linguists
approach the notion of markedness. The goal of this introductory
study is to render explicit both the ways in which the different
notions represented at the conference differ and also the ways in
which they nonetheless pertain to related issues. The following
section (section 2.) will discuss in more detail the common core of
the various markedness theories, trace this basic theme in the conference papers, and then identify the basic parameters along which
the theories differ. Section 3. will take up these parameters one
by one and place the papers within a general framework by making the
similarities and differences among the various concepts explicit.

2. The core notion

The core hypothesis of markedness theory pertains to correlations. The domain of the theory is in all cases pairs of opposing
language-structural entities that exhibit an asymmetrical relationship in more than one respect. The central claim is that the various tests that demonstrate the asymmetry between the two members of
the opposition will have converging results: Once one of the two
members has been shown to be marked by one criterion - let us say,
it has been shown to be structurally more complex than the other, or
paradigmatically poorer, or more restricted in its distribution -
all other relevant tests will also converge to select that entity as
the marked member of the opposition.

The empirical nature of the core hypothesis of markedness theory

stems from the fact that the tests that are claimed to show such <u>con-</u>
<u>verging</u> results about markedness assignment are themselves <u>logically</u>
<u>independent</u>. Thus, there is no logically compelling reason why a
more widely distributed object should be simpler or that it should
be more richly elaborated in terms of subtypes than a less widely
distributed object is. The theory thus exhibits the basic characteristic of significant scientific hypotheses: where plain logic
would posit no relationship among things, the theory predicts one.
Or, putting it differently, it claims things to be related which
otherwise would be seen as unrelated.

    The centrality of correlations to markedness theories of any
kind is documented by the fact that this notion is prominent in some
form in the markedness concept of most if not all papers at the conference.

    The fact that all versions of markedness theory incorporate a
claim about correlating properties of some language-structural entities still leaves much room for variability among versions of the
theory. A theory would constitute a <u>minimal</u> version of markedness
theory if it claimed that there is at least <u>one</u> language in which
there is at least <u>one</u> opposition type whose members exhibit a markedness relationship by the converging verdicts of at least <u>two</u>
tests. At the other extreme, a <u>maximal</u> markedness theory would make
a claim for <u>all</u> languages and it would hypothesize that <u>all</u> tests of
distribution, syntagmatic complexity and paradigmatic variability
yield converging results on the assignment of these entities to the
two poles of the markedness dimension. It is now easy to see that
in between these two extreme formulations there is a broad range of
possible variety concerning the empirical claims that markedness
theorists may choose to propose.

    There appear to be four principal dimensions to this broad varia-
tion. One has to do with the <u>type of entities</u> between which markedness relations are claimed to hold. Thus, for example, one theory
may claim that markedness governs phonological oppositions but not
syntactic ones while another may propose that it is a comprehensive
principle governing both areas. Second, although all tests distinguishing between marked and unmarked terms have to do with one of
the three basic types mentioned above, theories may nonetheless
differ with regard to the specific <u>criteria</u> that they hold to be
diagnostic of markedness relations. Third, markedness claims may
also differ with respect to the <u>domains</u> for which they claim markedness relations to hold. For example, some versions of markedness
theory may claim markedness assignments to be universal while others
may posit that their validity is bound by individual languages or
classes thereof.

    Finally, theories may also differ in how they construe the <u>signi-</u>
<u>ficance of markedness theory</u> within the context of a comprehensive

account of language structure. They may view markedness as a low-
level theory which is in need of further explanation within linguistics, or they may see it as an explanatory theory of the highest
kind possible within the boundaries of the discipline.

In what follows, we will consider the range of different claims
possible for markedness theories ordered along these four parameters
in more detail and will attempt to characterize the differences and
similarities among the conference papers by stating the position
they represent relative to the main issues.

3. Variables

3.1. Types of entities

Although the classic proposal concerning markedness was made by
Trubetskoy in relation to phonology, the notion has been richly explored by now with respect to all other aspects of natural language
grammars as well. Of the conference papers, six focus primarily on
phonology (Benson, Cairns, Cox, Fellbaum, Menn, and Scriven) and
nine primarily on morphology and syntax (Chao, Comrie, Gundel-
Houlihan-Sanders, Harbert, Lapointe, Odlin, Shauyman, Solan, and
White).

Markedness relations are at times claimed to hold between linguistic objects that are relatively concrete such as individual utterances or sets of utterances comprising entire languages. More
commonly, the terms of proposed markedness relations are more abstract entities: they are grammatical phenomena such as phonetic
segments or syntactic constituent types or relativization patterns.
Markedness relations may also be claimed to hold for particular
pairs of such phenomena (e.g. meaning-form relations) or larger sets
thereof such as entire phonologies of languages or their entire grammars.

Decisions one way or another as to the types of terms for which
markedness relations can properly be claimed are usually made in a
tacit fashion, rather than highlighted as issues. There is, however,
pertinent discussion in two of the papers included in this volume.
Cairns, in his study on phonotactic constraints of words, emphasizes
that markedness is primarily a property of grammars, rather than of
utterances or parts thereof. And Cox, dealing with the relationship
between palatalized consonants and vowel systems, makes the point
that a vowel system as a whole may be marked because of the asymmetric arrangement of the vowels included but without any of the individual vowels being marked when taken in isolation. A criterion for
judging entire grammars - or entire languages - as more or less
marked is offered by Lapointe. His Fundamental Markedness Principle Relating Forms and Notions proposes that a grammar is less
marked if less complex morpho-syntactic forms are paired with less
complex meanings.

3.2 Criteria

The question that is at the heart of any markedness claim con-
cerns the diagnostics of markedness relations. Given two terms, what
has to be true for them in order for us to be able to claim that
there is a relationship of markedness between them and, further, what
does it take to show that one, rather than the other, is the marked
member of the pair? The following chart identifies the various views
on this as we see them represented in the conference papers.

AUTHOR	CRITERIA ASSUMED OR ARGUED FOR:	CRITERIA REJECTED:
Benson	typological implication	
Cairns	syncretization	overt marking
	neutralization	frequency
	typological implication	
Chao	context-dependent interpretation	
	immediacy of interpretation	
Comrie	overt marking	
	expected meaning	
Cox	frequency across languages	
Fellbaum	typological implication	
Gundel-Houlihan-Sanders	typological implication	
	language-internal distribution	
Harbert	frequency across languages	
	direction of diachronic change	
Lapointe	frequency across languages	
	iconicity	
Menn	frequency	
Odlin	semantic dependence	overt marking
	restriction of usage	
	semantic range	
	range of derivational options	
	frequency	
Shauyman	language-internal distribution	
Scriven	frequency across languages	

Solan   availability to child

White   frequency across languages
              learnability

None of the criteria assumed or argued for in the papers are new;[2] the thrust of the papers has not been to propose new diagnostic devices for markedness but to test ones already proposed in the literature for particular domains of facts. Some of the papers attempt to reduce the inventory of markedness criteria by showing that some of them are logically implied by others. Thus Cairns proposes that increased frequency is a logically necessary result of neutralization. Although reducing the set of criteria by establishing causal links among them is difficult,[3] some insight into the interrelationship among the criteria results if we note the similarities that obtain among some of them. Indeed, as was suggested in the beginning of this paper, most if not all the criteria pertain to one of the three basic domains of syntagmatic complexity, paradigmatic complexity, and distribution.

Syntagmatic complexity has to do with the amount of structure that a linguistic object has. Of the criteria mentioned in the conference papers, two belong in this type: overt marking adopted by Comrie and argued against by Cairns and Odlin and semantic dependence used by Odlin. Overt marking refers to the presence of additional morphological or phonetic material: authoress has an overt marking as opposed to author, or nasal vowels may be said to be overtly marked over oral ones. Semantic dependence is a term Odlin borrows from Marchand; what it refers to is that of two words the meaning of one may "lean on" that of the other in the sense that the first includes all of the meaning of the first plus an additional bit of meaning. Thus, the verb to knife has a meaning that can be said to depend on the meaning of the noun knife since the former includes the latter plus verbality.

Paradigmatic complexity has to do with the number of subdistinctions available within a particular category. Of the criteria used in the papers, syncretization, discussed by Cairns and range of derivational options, discussed by Odlin, belong into this type.

Most of the criteria utilized in the papers belong to the third class: they have to do with distribution. Both typological implication and frequency across languages have to do with crosslinguistic distribution, the claim being that unmarked terms are more widely occurrent across languages than unmarked ones. Harbert's criterion of the directionality of diachronic change also appears to be a kin notion. Criteria that have to do with distribution within a language are neutralization (Cairns), context-dependent interpretation (Chao), restriction of usage and semantic range (Odlin), frequency within language (Menn, Shaumyan), and the various specific tests aimed at

tapping language-internal distribution as proposed in the papers by
Shaumyan and by Gundel-Houlihan-Sanders. This leaves only criteria
that do not readily fit into any of these classes: immediacy of
interpretation (Chao), availability to the child (Solan), learnability (White), and iconicity (Lapointe). The first three of these
are psychological, rather than structural-distributional, notions:
they have to do with a psychological interpretation of markedness
rather than with the linguistic notion proper.

The fourth, Lapointe's Fundamental Markedness Principle Relating
Forms and Notions which we referred to in the chart under the term
iconicity says in effect that grammars are less marked if simple
forms are paired with simple meanings and complex forms with complex
meanings. This markedness criterion differs in its logic from the
others proposed in that it makes markedness dependent on the simultaneous truth of a more than one condition. It says that a grammar
is less marked if both simple meanings are expressed by simple forms
and also complex meanings are expressed by comple forms. This markedness criterion therefore does not simply resolve into a single
parameter such as syntactic or paradigmatic complexity or distribution.

Beyond the mere choice of criteria, there is room for opposing
assumptions to be made also as to the logical nature of the claims
made with respect to the criteria. Some of the relevant choices are
the following:

a/ Is it to be claimed that the marked and unmarked terms must
differ with respect to any criteria in a given direction; or is it
to be claimed that they must not differ in the converse direction?
Take, for example, overt marking. Cairns and Odlin argue that in
some instances both marked and unmarked terms are equally supplied
with an overt marker or both are equally void of one. Whereas they
conclude that for this reason the criterion of overt marking is not
a reliable one for markedness, one might propose that they are nonetheless relevant criteria because it is never the case that the term
known independently to be the marked one has no overt marking while
the one known independently to be unmarked has an overt marking.
Gundel, Houlihan, and Sanders do in fact conclude that at least some
of the markedness criteria must be interpreted in this weaker
manner.

b/ Are markedness criteria connected by mutual or unidirectional
implications? For example, can we predict relativ text frequency
from relative complexity of form as reliably as we can predict
complexity of form from text frequency? There is no explicit discussion of this in the papers.

c/ Are the correlations among the criteria absolute or probabilistic? Menn argues strongly that the correlations must be declared

probabilistic only. <u>Comrie</u> also acknowledges exceptions.
    d/ Are markedness relations <u>binary or scalar</u>? They could turn
out to be scalar, rather than binary, in three ways:

    aa/ If something were found to be marked by only some of the re-
levant criteria but not by others. For discussion, see the papers
by <u>Cairns</u> and <u>Odlin</u>.

    bb/ If any one criterion could apply to structures to lesser and
higher degrees. It is in this sense that <u>Odlin</u> argues for markedness
being scalar: he says denominal verbs in English are more marked vis-
a-vis their stems than deadjectival ones.

    cc/ If the same term could be shown to be marked relative to one
structure but unmarked relative to another so that a chain of markedness could be established. Such chains are proposed in the papers
by <u>Benson</u> and <u>Lapointe</u>.

    3.3. <u>Domains</u>

    How restricted are markedness claims in terms of the context for
which they are claimed to hold? <u>Cairns</u> argues that one cannot make
uniform marked-unmarked assignments in phonology about segments
unless one considers the phonetic contexts in which segments occur.
Are markedness assignments restricted in terms of the range of languages for which they hold? In principle, it is of course conceivable that identical pairs of terms show variable markedness polariza-
tion in different languages: the assignments do not need to be universal. While this question did not receive explicit discussion in
the papers, a related question did. Assuming that markedness assignments are universal among human languages, just what forms of human
communication do they hold for? Everybody would agree that they
hold for "primary" languages such as English and German and Swahili
as spoken by healthy adults. Do the same markedness relations
obtain in secondary language forms also such as child language or interlanguage, i.e. the language of the language learner? <u>Lapointe</u>,
<u>Menn</u>, <u>and</u> <u>Solan</u> deal with child language data, <u>Benson</u>, <u>Fellbaum</u> <u>and</u>
<u>White</u>, with interlanguage evidence.

    <u>Cairns</u> argues that markedness assignments may differ whether one
considers production or perception. And one could raise the question
of whether markedness assignments hold in linguistic competence or
linguistic performance.

    <u>Comrie</u>'s paper has an interesting implication for an even wider
application of the notions marked and unmarked. His claim is that
unmarked forms express expected meanings and marked forms stand for
less expected meanings. The notions "expected meaning" and "less
expected meaning" in turn correspond to situations in human experi-

ence that are more frequent and more natural. Thus, he implicitly
suggests – as we did in the beginning of this paper – that these
notions have applicability beyond language.

3.4 Significance

Three questions arise concerning the overall significance of markedness theories:

a/ Is markedness theory crucially linked with any other theory
of language or can it be independently maintained regardless of
other assumptions about language structure? In other words, to what
extent is markedness theory-bound or theory-independent?

b/ What, if anything, does markedness theory explain?

c/ What explains markedness theory?

Regarding the first question: most of the papers presented at
the conference define the terms of markedness relations in relatively
theory-free terms. Two of the papers, however – those by Harbert
and Solan – assume Government and Binding Theory as their frame of
reference.

As far as the explanatory significance of markedness is concerned, different views are represented in the papers. Menn argues
that markedness does not explain anything while Harbert, Lapointe,
and Solan see markedness as explanatory.

Third, what explains markedness relations? Lapointe says markedness relations follow from acquisition. Gundel-Houlihan-Sanders attempt a psychological explanation. Comrie proposes that markedness
in language is explained by markedness in the real world.

Whereas perhaps all would agree that markedness theory provides
generalizations from which particular facts follow – whether this is
considered to be explaining those facts or not – the need for explanations of markedness relations themselves continues to remain an
important challenge.

4. Conclusions

Although there is a core set of parameters within which concepts
of markedness currently understood and used generally lie, there is
considerable variation in precisely how markedness is used and understood. Claims that one linguistic structure is more marked than another one range over a wide domain of diverse phenomena. Thus markedness theory – if theory is the correct word – is really a family
of hypotheses: if one correlation is not borne out, others may still

hold. The multiplicity of empirical correlations among linguistic phenomena that can be generated by any markedness claim engenders - indeed demands - future research on the precise nature of the asymmetric oppositions in terms of which people construe and construct their linguistic world.

Footnotes

1. One of the few languages that have no voiceless stops may be the Australian language Gudandji which, according to Aguas has b̲, d̲, and g̲ but no p̲, t̲, or k̲. The voiced stops b̲, d̲, and g̲ are, however, described as "lightly voiced" (Aguas 1968, 1f).

2. For surveys of markedness criteria, cp. Greenberg 1966, Brown and Witkowski 1980, Mayerthaler 1982.

3. For a thought-provoking attempt to establish causal relations among the classic markedness criteria in Greenberg 1966, see Schwartz 1980.

References

Aguas, E. F. 1968. Gudandji. Pacific Linguistics, Series A: Occasional Papers 14. 1-20.

Greenberg, Joseph H. 1966. Language universals. The Hague: Mouton.

Brown, Cecil H. and Stanley R. Witkowski, 1980. Language universals. In David Levinson and Martin J. Malone, ed., Toward explaining human culture: a critical review of the findings of world wide cross-cultural research. HRAF Press. 359-384.

Mayerthaler, Willi, 1982. Markiertheit in der Phonologie. Unpublished.

Witkowski, Stanley R. and Cecil H. Brown. 1983. Marking-reversals and cultural importance. Language. 59, 3, 569-582.

Schwartz, Linda J. 1980. Syntactic markedness and frequency of occurrence. In Thomas Perry, ed., Evidence and Argumentation in Linguistics. Berlin: Walter de Gruyter.

WORD STRUCTURE, MARKEDNESS, AND APPLIED LINGUISTICS

Charles E. Cairns

Queens College, CUNY, and

Doctoral Program in Linguistics, CUNY

This paper describes a theory of word structure which incorporates aspects of markedness theory. The theory is illustrated by an account of some facts of English phonotactics and of first and second language acquisition. The phonotactic phenomena serve as the basis for justifying a rich hierarchical structure of English words. The theory presented here is an extension and revision of the one described in Cairns (1982), where I justified a structure of English monomorphemic words which contains syllables ($) and feet (F) as labelled nodes subordinate to the root node word (W). With the exception of some word-initial syllables, all syllables are in bisyllabic or (under special conditions) trisyllabic feet. The first syllable of the English foot is labelled the head syllable ($_h$). $_h$ enjoys special status in two ways: It is the repository of relative prominence (stress) and allows a greater variety of phonemes and combinations thereof than nonhead syllables. The theory of word structure is illustrated in this paper by an account of aspects of English consonant clusters, especially in the onset; the overall structure of the syllable is also a key focus, especially in the illustrations of the application of this approach to applied linguistics.

Lexical phonotactic constraints are accounted for by a minimal specification of distinctive features, "deposited" at labelled prosodic nodes, plus universal conventions for interpreting such entities as strings of phonemes. Among the more important aspects of phonological redundancy accounted for in the present theory is that which inheres in the order of phonemes. For example, since stops may not precede fricatives in English onsets, a redundancy-free representation of a two-obstruent onset should contain no ordering specifications. Instead, the order fricative-stop should follow automatically from a lexical representation where the features of the stop and of the fricative inhere in a higher level node.

An important feature of the general approach to phonology followed here is the belief that much can be learned from a shift of focus away from rules to representations. This accounts for the great importance paid to phonotactic constraints within the lexicon; it is within this "static" domain that we can get a picture of prosodic structure most uncomplicated by the operation of phonological or morphological rules.

The distinction between rules and representations--and the particular importance of focussing on the latter--has been employed before in studies of applied linguistics. Eckman (1984), for example, observes that phonotactic constraints in interlanguage obey universal properties of language, whereas rules do not. In a similar vein, Lise Menn (1978) makes the same distinction in her studies of child phonology. In the theory described in this paper, all the phenomena which Eckman and Menn describe by means of phonological rules (i.e., phenomena which seem to threaten the prospect that all aspects of interlanguage or child phonology obey universal constraints) are handled by universal conventions applying to lexical entries.

In the following section of this paper I attempt a definition of "markedness" as used in this paper and contrast it with other uses of the term as it has been used in the literature and in this conference. This is followed, in Section 2, by an exposition of my theory of syllable structure. Section 3 describes the application to second language acquisition, and section 4 contains an application to child phonology.

1. MARKEDNESS AS A PROPERTY OF GRAMMARS

The term "markedness" has been used in many ways in the linguistic literature. Thus, Rutherford (1982) equates markedness both with (putative) psycholinguistic complexity and with notions of sentential complexity. More commonly, the term has been used to refer to some aspect of linguistic complexity. The general idea is that if a particular linguistic feature is marked, it is somehow unexpected or adds a degree of linguistic complexity. The usual approach (and the one adopted here) is to look for purely linguistic evidence in order to determine what is and what is not marked. Thus, such linguistic phenomena as syncretization, neutralization and, especially, implicational universals are typically taken as valid indicators of markedness. Two other phenomena which are frequently (and in my view mistakenly) confounded with markedness are the presence vs. absence of overt markings and various kinds of frequency counts.

In the paragraphs which follow I discuss these various proposals for determining what is and what is not marked. But it is first worthwhile to give my general view of markedness as primarily a property of grammars, as opposed to a property of utterances or parts of them. In particular, I define markedness in terms of universal parameter theory. According to this theory, universal grammar contains a set of discrete parameters, and a grammar may be constructed by going along some of them (there may be restrictions on the combinations of parameters which may be chosen); each parameter consists of a set of discrete points, where each point represents a marked feature, so that the grammar of a language may be described in terms of the total configuration of M's it has; since each M is a point on a parameter, the total configuration of M's reveals which parameters have been chosen and which point on each parameter. This is exemplified in section 1.6, below. In the intervening five sections, I contrast this general perspective with other views of markedness, all of which have played a role in the presentations at this conference.

1.1 Overt markings. A common view of markedness in both phonology and morphology is to equate the concept with the presence vs. the absence of a marker of some kind. In phonology, for example, glottalization in obstruents is marked because it is characterized by the presence of an "added on" phonetic feature. The regular English plural would be an example of markedness in morphology, because it is indicated by the presence of a morpheme.

The concept stems from Trubetzkoy's concept of privative oppositions, those characterized by the presence vs. absence of a feature (Trubetzkoy, 1938 ms., 1958). The unmarked member of a privative opposition--the member without the added overt marking--is also, according to Trubetzkoy, the member which is supposed to appear in the position of neutralization (cf. Section 1.4, below). According to Trubetzkoy's scheme, nonprivative oppositions (e.g., tongue height in vowels, point of articulation in consonants) were not definable in terms of markedness. Jakobson extended the concept to morphology in a 1939 article entitled "Signe Zéro," where the addition of a grammatical marker is made equivalent to making a form more marked. Let us call this general view the overt marking theory of markedness.

The overt marking approach to markedness is deficient in both phonology and morphology. In phonology, one of the main problems is the determination of the level at which the supposedly marked feature is considered to be added. Is it at the level of physical phonetics or at a level of linguistic theory, properly defined? Consider the example of nasality--the velum is normally open when the vocal tract is at rest, and a positive gesture is required to close it for nonnasal speech sounds. However, linguistic criteria (as discussed in sections 1.3 - 1.5, below) classify nasality as marked in vowels.

For another phonological example, consider lip rounding in vowels. On linguistic grounds, the presence of rounding is marked for front and low vowels but unmarked for vowels which are both back and nonlow. Articulatorily, the lip rounding is essentially the same in both the marked and the unmarked vowels, so there is no correspondence between linguistic markedness and the presence or absence of an articulatory gesture. The fact that perceptual arguments have often been evinced to handle this example raises the question of deciding when articulatory or perceptual arguments are appropriate.

It is always possible to interpret some subset of the entire set of articulatory gestures associated with any given sound as representative of an added gesture if the analyst wishes to consider the sound marked. Analogous moves could be made in the perceptual domain, thus revealing the essential arbitrariness of an overt marking approach in phonology.

A compelling refutation of the overt marking theory of markedness in morphology follows from the existence of zero-derivation. Odlin (this volume) argues that in English, zero-derived denominal verbs e.g., to stone, to bridge, etc.) are marked with respect to zero-derivation of verbs from adjectives and statives in general. If any set of zero-derived forms can be said to be morphologically marked, then we have a case of markedness in the absence of overt markings.

Just as a form can be marked in the absence of a marker, unmarked forms may occur despite the presence of a marker. Consider, for example, the unmarked status of the masculine gender in Spanish. Greenberg (1966) argues convincingly (on grounds of neutralization cf. Sec. 1.4) that masculine is unmarked, despite the fact that it is usually overtly indicated by a morpheme.

Spanish indicates gender with /-o/ and /-a/ (masc. and fem. respectively), so el hijo and la hija mean 'son' and 'daughter,' respectively. The unmarked nature of the masculine gender is shown by the usage los hijos, which may refer to a mixed group of sons and daughters. Thus, when gender is neutralized by extralinguistic context, the masculine appears. A further example is the phrase el hijo y la hija son buenos, 'the son and the daughter are good' where the adjective bueno has the masculine morpheme. Here again, when the masculine vs. feminine distinction is neutralized, the masculine appears. If we accept the notion that feminine is a marked gender with respect to masculine, then we have an example of markedness where the unmarked members of the morphological opposition is signified by an overt marker, just as is the marked member.

To summarize the discussion of the overt marking theory of markedness, it appears that it is fraught with difficulties of numerous types in both phonology and morphology. At the heart of all these problems lies an assumption that there are some acceptable means for determining markedness; it is argued later that only linguistic criteria should be employed. In many cases, entities which linguistic theory tells us should be marked are not clearly characterized by the presence of an overt marker, and there are other cases of unmarked forms with an overt marker.

1.2 Frequency. Trubetzkoy noted an uncanny relationship between those members of phonological oppositions he thought of as unmarked and their higher lexical and text frequency. Greenberg used this relationship as a major argument in his book Language Universals (1966). Ferguson (1963) argues that there are always fewer nasal segments in a language's inventory than there are oral cognates; furthermore, text frequency counts performed by Greenberg and others show a much higher frequency for oral than for nasal vowels. Greenberg argues that those entities which linguistic criteria classify as unmarked frequently have a higher text frequency than the unmarked; furthermore, in inventories, the marked rarely (if ever) outnumber their unmarked cognates.

Despite the commonly occurring relationship between frequency and markedness, it is a mistake to use frequency as a definition of markedness. A particular feature may show a high text frequency for factors having nothing to do with markedness. For example, Icelandic has a rule which inserts a front, rounded [ü] in the environment: C___r $\left\{ C, \# \right\}$, a commonly occurring environment. Obviously, all features of this segment are redundant, but the common presence of a front rounded vowel would throw off counts of the text frequency of front rounded vs. front unrounded vowels; [ü] cannot be marked for any of the features it possesses because none of its features are distinctive—the entire segment is redundant.

Historical factors can also interfere with a perfect match between statistical counts and markedness. For example, some languages have more long than short mid vowels (e.g., Karok, some dialects of Arabic), despite the fact that linguistic criteria favor analyzing length as marked. Greenberg (1966, p. 22) suggests that the historical tendency for ai and au to monophthongize to e: and o:, respectively, may explain this discrepancy (see Cairns 1977 for further discussion).

The conclusions to be drawn from statistical studies is that markedness is so loosely correlated with frequency trends that it is at best a highly unreliable guide in determining markedness.

1.3 Syncretization. Syncretization, along with neutralization and implicational universals, is more properly a feature of synchronic linguistic structure than are any sorts of frequency counts; accordingly, it is at least potentially of more interest in deter-

mining markedness assignments. The essence of the relationship
between syncretization and markedness lies in a tendency to avoid a
proliferation of marked features in any one entity. Thus, in a
marked member of any given opposition, fewer other oppositions are
likely to occur, because, for each further opposition, there would
have to be a marked member.

To give an example from Spanish morphology, let us assume that
the subjunctive is marked with respect to the indicative. The
tenses have collapsed to two in the subjunctive, present and
imperfect, whereas the indicative mood has the preterite and future
tenses as well. The coding tree in Figure 1 illustrates the point
that if the Subjunctive mood were to allow further subdivisions
within the present and imperfect, there would appear more forms
which would be highly marked.

(1)

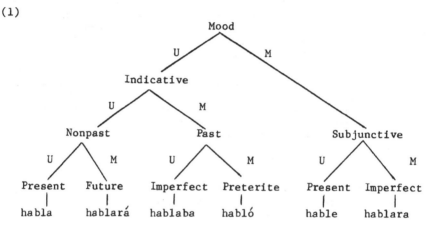

Figure 1.

For a phonological example, consider the Finnish front vowels:
/i/, /ü/, /e/, /ö/, and /æ/. The rounded vowels, being marked, do
not show the distinction between low and non-low vowels; accordingly, this distinction is syncretized in this marked category.

Syncretization is more properly a linguistic feature than are
statistical tendencies, because syncretization involves the presence
or absence of linguistic distinctions in certain subcatagories of
the lexicon. Markedness is, of course, defined within linguistic
theory and phenomena such as syncretization, neutralization, and
implicational universals are among the evidence for arguing in favor
of specific markedness assignments. Of the three, syncretization is
somewhat less compelling than the other two, since it is not as
frequently associated with markedness.

1.4 <u>Neutralization</u>. Neutralization is similar to syncretization and may account for some of the frequency phenomena mentioned above. This is so because if it is the case that the unmarked member of an opposition is the only one destined to appear in a position of neutralization, then the unmarked member of the opposition will have more opportunities to appear in a text and thus may have greater text frequency. (This makes frequency counts appear superfluous, in addition to potentially misleading.) Furthermore, syncretization may be viewed as neutralization, where the context is defined in terms of simultaneous instead of neighboring properties.

According to Trubetzkoy, neutralization is a defining characteristic of markedness. As such, it could apply only to privative oppositions. All contemporary approaches to markedness in phonology go well beyond Trubetzkoy's conception (cf. Cairns, 1971). In particular, markedness is seen as applying to all phonological oppositions, the distinction between privative and nonprivative oppositions no longer considered valid in phonology. Furthermore, all linguists insist on the universality of markedness assignments, whereas Trubetzkoy allowed for markedness to be determined by language-particular criteria (such as neutralization).

Markedness is properly seen as a construct of phonological theory, such that neutralization, implicational universals and syncretization constitute evidence in favor of or against specific markedness assignments. In the theory presented in this paper, the sub-classes of the entire phoneme inventory which are allowed in each position of the syllable are defined. Within each sub-class, the phonemes are portrayed as opposed to each other by features, each of which has a marked and unmarked value. Neutralization rules account for most of the phonotactic constraints by eliminating the possible occurrence of marked feature values in specific environmental contexts.

1.5 <u>Implication universals</u>. Implicational universals have traditionally been assumed to be the most reliable index of what is and what is not marked, and with good reason. Thus, in morphology, it is observed that many languages have overt markings for plural, some have overt markings for both singular and plural, but none have a morpheme indicating singular with no indication for plural. Accordingly, plural is taken as the marked number. The fact that many languages show syncretization in the plural but none do in the singular (e.g., German, where gender distinctions are neutralized in the plural) illustrates the importance of supporting markedness claims from syncretization and neutralization with those from implicational universals. In the theory presented in section 2, below, implicational universals are the basis for markedness assignments (although space limitations preclude a justification in this paper).

1.6 <u>Parameter theory and syllable structure</u>. The parameter
theory approach to markedness sketched in the introduction to this
section is related to the above considerations as theory is to data;
syncretization, occurrences in positions of neutralization, and
implicational universals are all evidence on the basis of which the
linguist determines parameters and points along these parameters.
In the following exposition of syllable structure, implicational
universals will be the key evidence for determining the parameters;
neutralization will be introduced as a formal device, part of an
apparatus designed to account for phonotactic constraints within
syllables.

Consider (2), adapted from Cairns and Feinstein (1982) and based
on Greenberg (1978). The implicational universals are depicted in
(3); thus, all languages contain at least type (2a), no language may
have type (2c) without also containing type (2b), and the choice of
type (2d) is independent of either (2b) or (2c).

(2)

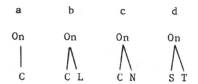

On = Onset; C = obstruent; L = liquid
or glide; S = fricative; T = stop; N = nasal.

(3)

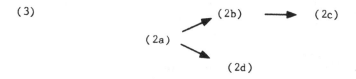

Each of two branches in (3) represent a different parameter,
with the beginning, least marked point depicted by (2a). If a
language complicates its onset system, it may do so in either or
both of the two ways depicted here. (There are, of course, other
alternatives, not discussed here.) If the upper parameter is
chosen, then the first marked point would be type (2b), the second
(2c). A full elaboration of this approach would reveal a large
array of parameters and sub-parameters, with some constraints on
choices among them; the above suffices as a sketch of the approach
taken here.

2.   A THEORY OF SYLLABLE STRUCTURE

Phonotactic constraints within the syllable are accounted for by
means of a phrase marker for syllables, complete with labelled
nodes, along with a specification of the classes of phonemes allowed
in each position of the syllable. This system is supplemented by a
set of neutralization rules, alluded to in section 1.4. Phonotactic
constraints across syllable boundaries requires the invocation of
feet and other supra-syllabic structures. In this paper, I provide
an overview of syllable structure, with emphasis on the onset.

This discussion is confined to stressed syllables, since these
allow for the greatest variety of phoneme combinations. Unstressed
syllables contain a proper subset of the consonants and consonant
clusters allowed in stressed syllables. It should be noted that
stress is accounted for in a basically passive way; stress is
directly derivable from the prosodic structures resulting from
application of the conventions portrayed below to lexical
representations.

2.1 <u>Constituent structure of the English syllable.</u> The
maximally long syllable of English permitted under my theory is
depicted in Figure (4), for the word <u>sprinkle.</u>

(4)

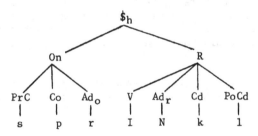

The term PrC abbreviates PreCore; Co = Core; Ad_o = Adjunct in the
onset; Ad_r = Adjunct in rhyme; Cd = coda; PoCd = PostCoda; R =
rhyme; $\$_h$ = head syllable; V = vowel.

The analysis of the final /l/ as within the same syllable requires justification. The primary (in fact, for our present purposes, the <u>only</u>) reason for making syllable divisions is to arrive
at the most parsimonious account of phonotactics. (Articulatory
and perceptual syllable divisions may be quite distinct from those
flowing from phonotactic considerations.) Observe that there are
co-occurrence restrictions between the final /l/ and the make-up of
the Ad_o constituent. An /l/ may not appear both in the Ad_o and
anywhere else within the syllable. English has words like <u>sprinkle,</u>

splinter, crystal, but there are no *splinkle, *splintle, *clystal.
(Similarly, we have words like lull, where the initial /l/ is the
Onset Core constituent, but no *clull.) Since we assume that all
such co-occurrence restrictions are best handled tautosyllabically,
parsimony dictates analyzing sprinkle as a monosyllabic word. (See
Cairns 1982 for further elaboration.)

I assume a universal stock of labels, and phrase structure rules
drawn from a universal stock which generate syllables. The relevant
rules are depicted in (5).

(5)

$$\$_h \longrightarrow (On) \quad R$$

$$On \longrightarrow (PrC) \quad Co \quad (Ad_o)$$

$$R \longrightarrow V \quad (Ad_r) \quad Cd \quad (PoCd)$$

I will turn to the question of the role of these structures in
the theory of phonology below, but for now take them as mechanisms
for defining syllabic templates. Observe that my analysis has
English stressed syllables with an optional onset, but an obligatory
coda. The optionality of the onset is obvious, but the obligatory
nature of the coda requires justification. For one thing, I assume
the glide portion of what is usually referred to as tense vowels to
be a member of one of the constituents in the rhyme other than the
'V' constituent. (This would include words like spa, which takes a
centering glide.)

In (6) I provide analyses of several words with vowel-glide
sequences, and one with a lone vowel followed by the /sp/ cluster,
as in the word lisp.

(6)

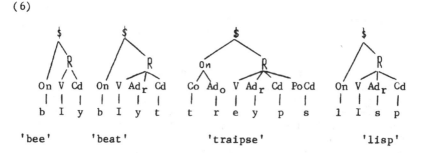

'bee'   'beat'   'traipse'   'lisp'

It follows from my analysis of glides as coda elements that the coda is obligatory in word final, stressed syllables, because no such syllables may end in a lax vowel. The glide or consonant which follows the vowel is in one of the constituents after the V. If there is only a glide, it occupies the Coda constituent as in bee. If there is both a glide and a single obstruent, the glide occupies the Ad$_r$ constituent, and the obstruent occupies the coda slot, as indicated in the diagram for beat. And if there is a vowel followed by a glide and two subsequent consonants, as in traipse, then the last consonant occupies the post-coda position.

The example discussed above illustrates how tense vowels as well as lax vowels followed by consonants are treated in my approach. Since no English words end in a stressed, lax vowel, it appears tempting to claim that all stressed syllables must have a coda. This would entail, however, analyzing single consonants which appear between a stressed vowel (to its left) and an unstressed vowel as the coda of the left-hand syllable as opposed to the onset of the right syllable, contrary to most current practice. In Cairns 1982 I defend this analysis and take it further, claiming that intervocalic consonant clusters are normally analyzed as the coda of the left syllable (provided the right syllable is without stress). The argument in favor of this approach is, again, parsimony in accounting for phonotactics. Such intervocalic clusters are more neatly accounted for by classifying them as codas than as onsets or as sequences of codas followed by onsets. The only exceptions are a few borrowings like vodka and some morphological relics like pamphlet.

2.2 Phonotactic constraints. Phonotactic constraints are accounted for by defining the classes of phonemes which may occupy each of these labelled positions. For illustration of this concept, observe the analysis of lisp in (6). In Cairns (1982) I argue that /p/ is not allowed in PoCd, but /s/ is allowed in Ad$_r$ position. Thus, there is only one analysis of lisp, as shown. A glide can not appear before an /sp/ cluster in the same syllable because then the glide would occupy the Ad$_r$ slot, the /s/ would be the coda, and the /p/ would be in the PoCd position, in violation of a rule of English. Accordingly, the nonoccurrence of a word like */liysp/ is accounted for, but the sequence in traipse is allowed.

2.3 Lexical and phonological representation of the syllable. The syllable is represented differently at different phonological levels. There are three levels of interest--the (systematic) phonetic, the phonological and the lexical. The phonetic level, where all linguistically significant phonetic information is fully specified (e.g., aspiration in English, etc.), is not discussed in this paper. The representations in (6) correspond to the phono- ·
logical level, defined as the level at which inflectional morphology and a variety of phonological rules apply.

Lexical representations do not contain any redundant information. Accordingly, the initial /s/ in an onset cluster such as
sp-, for example, need not have the features of stridency and
voicelessness specified, since these are redundant features.
It has been less frequently noted that there are significant
redundancies to be captured in the order of phonemes within constituents. Menn (1978) is among a few linguists to observe this
important fact. For example, if an English onset is specified to
contain both an /s/ and a /p/, the order is redundant because
English does not allow ps onsets. Therefore, the lexical representation of such an onset should contain only the distinctive features
necessary to specify the information that it contains the phonemes
/s/ and /p/, and no information about the order of these phonemes.
In the theory of phonology advocated here lexical representations
contain a minimal amount of structure; for example, the labels of
obligatory nodes are not specified. A set of conventions convert
lexical representations into forms like those in (6).

To see how the system works, consider the lexical representations of the words tap, pat, and apt in (7). Observe that in all
three, the only node labels mentioned are the optional On and Ad_r.
The capitalization of ONSET and AD_r means that these terms refer
to features, which trigger the assignment of the phonological features within the brackets labelled by this feature to the syllabic
constituent with the same label. (The peculiarities of the representation of apt are discussed separately below.)
(7)

$$
/t\text{æ}\,p/ = \begin{bmatrix} \text{ONSET} \\ +\ \text{cor} \\ -\ \text{voi} \end{bmatrix} \qquad /p\text{æ}\,t/ = \begin{bmatrix} \text{ONSET} \\ -\ \text{cont} \\ -\ \text{cor} \\ -\ \text{voi} \end{bmatrix} \qquad /\text{æ}\,pt/ = \begin{bmatrix} \text{AD}_r \\ -\ \text{cont} \end{bmatrix}
$$

$$
\begin{bmatrix} +\ \text{low} \\ -\ \text{bck} \\ -\ \text{cor} \\ -\ \text{cont} \\ -\ \text{voi} \end{bmatrix} \qquad\qquad \begin{bmatrix} +\ \text{low} \\ -\ \text{bck} \\ +\ \text{cor} \\ -\ \text{cont} \\ -\ \text{voi} \end{bmatrix} \qquad\qquad \begin{bmatrix} +\ \text{low} \\ -\ \text{bck} \end{bmatrix}
$$

The coding trees in (8) are part of the mechanism for deriving
the phonological representations from the lexical representations;
they also define the class of phonemes which occur in each of the
positions of the onset. The order of features in the hierarchy is
in part fixed by an attempt to capture the insights of Kean (1975),
but I have also employed some rather arbitrary decisions which can
be empirically investigated. The markedness assignments are on the
basis of arguments similar to those in Cairns and Feinstein (1982).
(Markedness is not employed in this part of the algorithm.) The
coding tree for the Core constituent contains some aspects which the
reader may find surprising—the absence of /s/ and the presence of

/kʷ/, for example. These and other apparent anomalies are explained in Cairns (1982).

(8)   Coding Trees for Onset Constituents

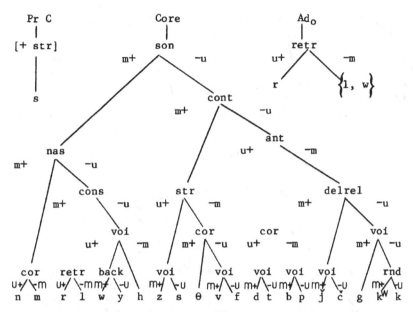

To derive the /t/ of tap, we look only at the Core constituent
in (8), because neither P_rC nor Ad_0 contains any of the features
specified in the lexical representation with the feature, namely
[+cor] and [-voi]. These ONSET specifications are the minimal
number of feature specifications necessary to uniquely specify /t/
in this coding tree. Specifications is accomplished by means of an
algorithm which traces a path, which may not double back upon
itself, starting from the root node, Core, and terminating at a
phoneme at the bottom of the tree. This path must contain the
features specified in the lexical representations. The most
parsimonious theory of English lexical phonology is consistent with
the structures in (8) (see Cairns, 1982).

The rhymes are specified in a similar manner, except here we do
not have node labels for the obligatory V and Coda. However, since
the features for consonants and vowels are different in my system,
the features in the "unlabelled" portions of (7) uniquely select the
vowel (V) and Coda constituents. The coding trees for the rhyme
constituents, some of which are provided in (9), are motivated in
Cairns (1982).

(9)   Coding Trees for Rhyme Constituents
       (excluding vowels)

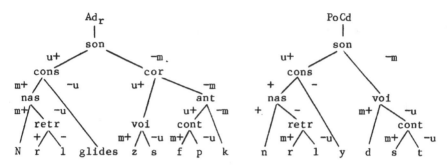

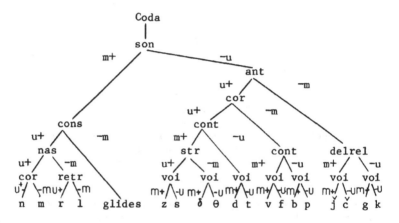

Notice that the same segment may be specified with quite
different features, depending on its constituency. For example, the
/t/ requires no features for its specification in <u>apt</u>, because /t/
is the only segment allowed in Coda position, according to the
analysis in Cairns (1982). This follows from the fact that final
<u>⁻pt</u> and <u>⁻kt</u> clusters are the only coda sequences consisting of /p/
or /k/ followed by another consonant which allow only lax vowels to
their left. Final <u>⁻ps</u> and <u>⁻ks</u> sequences, for example, allow a glide
or nasal in Ad_0 position, as in <u>traipse</u>, <u>hoax</u>, <u>glimpse</u>, and <u>lynx</u>.
Accordingly, these sequences must be analyzed with the /p/ or /k/
occupying the Coda position and the /s/ as a PoCd, allowing room for
the glide or nasal in the Ad_r slot. In words like <u>apt</u> and <u>act</u>,
however, the analysis of the /p/ and /k/ as an Ad_r accounts for
the absence of a glide or nasal in this position. Since <u>⁻pt</u> and <u>⁻kt</u>
are the only sequences which require this analysis, then it follows
that /t/ is the only segment which can appear as a Coda once either
/p/ or /k/ is chosen as the Ad_r segment, and is therefore totally
predictable.

Since /t/ is predictable in apt, it requires no features in the lexical representation, whereas features are required to specify /t/ in tap and pat, since it is distinctive in these positions.

There is not necessarily a unique specification for each lexical representation. Thus, /p/ in the Onset must be specified [-voi] and [-cor], but either [+ ant] or [-cont] may be used in the lexicon to distinguish /p/ from /f/. This follows from the fact that /p/ is distinctively [+ ant], in opposition to /k/, whereas /f/ is not distinct from any [-ant] fricative.

Observe that phoneme ordering is coded differently in this theory than in theories which assume lexical entries to be strings of phonemes. The basic order of consonants and vowels within the syllable is not represented at all in the lexical forms of any word, but is supplied by a set of PS rules which in turn obey tight universal constraints. The relative order of /p/ and /t/ in tap and pat, and of /p/ and the vowel in pat and apt, are coded by specifying node lables. This is just the right balance of specification, since those aspects of syllable structure subject to general constraints are not repeated for each lexical entry, whereas distinctive ordering distinctions receive extremely parsimonious, albeit indirect, representation.

The role of neutralization within this model is best explained by an analysis of the English Onset, to which we now turn.

2.3 The English onset. Redundancies in the number and order of segments in a syllabic constituent are illustrated in (10), where there is an ONSET with the features [-cont, - cor, - voi, + retr]. The algorithm sketched above reveals that there is no single path from the root of the Core tree to a terminal node which contains all these features; therefore, the next longer expansions of the ONSET node are attempted, shown in (11). The PrC - Core expansion is clearly not adequate, since [+ retr] is not contained in the PrC constituent. The Core - Ad_0 constituent, however, provides us with the representation we need; the features [-cont, - cor, - voi] uniquely specify the phoneme /p/ in the Core, and [+ retr] uniquely specifies /r/ in the Ad_0. There is no other ONSET segment or sequence of segments which can be represented by these features.

(10)

$$
\begin{bmatrix}
\text{\$} \\
\text{ONSET} \\
+ \text{ cont} \\
- \text{ cor} \\
- \text{ voi} \\
+ \text{ retr}
\end{bmatrix}
$$

(11)

There are obviously further constraints in the ONSET to be captured. For example, only a subclass of the Core phonemes may occur
if the Ad constituent appears. The complete list of English onsets
is given in (12). (Note the complementary distribution between /l/
and /w/ in the Ad_0 slot; this is discussed at length in Cairns
(1982).)

(12)   English Onset Clusters

        Core-Ad_0 Onsets
                         Core

    s   θ   f   d   t   b   p   g   k
  r šr θr fr dr tr br pr gr kr

Ad_0
$\begin{Bmatrix} 1 & w \end{Bmatrix}$ sl θw fl dw tw bl pl gl kl

            PrC-Core Onsets
                    Core
      n   m   w   y   t   p k^w k
  PrC sn sm sw š   st sp sk^w sk

            PrC-Core-Ad_0 Onsets
                    PrC-Core
              sp   st   sk
        r   spr   str   skr
Ad_0
$\begin{Bmatrix} 1, & w \end{Bmatrix}$ spl   ---   (skl)

Observe that in the Core-Ad_0 Onsets, the class of possible
Core phonemes can be described by clipping off the branches in
diagram (8) marked <u>M</u> for the features <u>sonorant</u>, <u>delrel</u>, <u>round</u>,
and those specifications of <u>voice</u> which are dominated by the
marked feature of <u>continuant</u>. Thus, I would propose the
neutralization rule in (13).

(13)

        N(son, delrel, round, <voice>)/$\begin{bmatrix} <+ \text{ cont}> \\ +/- \text{ retr} \\ [\text{ANY FEATURE}] \\ \text{ONSET} \end{bmatrix}$

This rule says that only the unmarked feature values of
sonorant, delrel, and round may appear in an ONSET which is also
specified for either '+' or '-' retroflex, plus one other
feature. (It is necessary to indicate that there must be one
other feature besides retroflex in order to allow for an /r/ or
/l/ to appear alone in an onset as the core element.) The
material in angled brackets indicates that if the ONSET is also
specified [+ cont], then the feature of voice is neutralized.

The rule in (13) has the effect of pruning the coding tree
in (8) to produce the coding tree in (14). This is the second
and last role for markedness in the formal theory of phonology
presented here to specify the branches of coding trees which are
pruned by neutralization rules. (Specification of parameter
points was the first.) Observe that the theory presented here
incorporates Trubetzkoy's suggestion that only the unmarked
member of an opposition may appear in a position of
neutralization.

(14)   Coding Tree for Core if Onset is [+/-retr],
                    plus another feature.

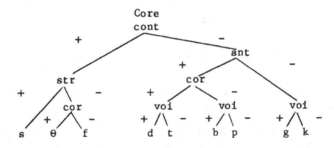

Turning now to the PrC-Core onsets, observe the neutralization rule in (15). This has the effect of pruning down the Core
tree to that shown in (16). If the onset is specified for both
the feature of stridency and plus or minus retroflex, we end up
with the tree in (17).

(15)   N(cont, delrel, cons, ⟨voice⟩)/

(16)

Coding tree for Core if Onset is
[+str] and any other feature except
voice.

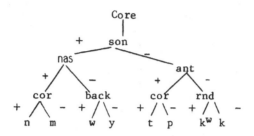

(17)

Coding tree for Core if
Onset is [+/- retr] and
[+ str].

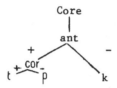

A more thorough analysis of the onset and rhyme in English is
provided in Cairns (1982). Since then, the theory has evolved
considerably; in the 1982 paper, the concept of eliminating the
phoneme strings entirely from the lexicon was not fully worked out.
A fuller exposition of the theory is not possible in this paper, in
order to allow space for describing its application to applied
linguistics. I have provided a complete list of the features
necessary for specifying all English onsets in (18) for any reader
who may wish to examine this theory more closely by working through
the algorithm for transforming lexical into phonological
representations.

(18)   Feature Analysis of all Complex
                   Syllabic Onsets of English

Core-Ad$_0$ Onsets   PrC-Core Onsets

sr/sl	=	[+str][+/-retr]	sn	=	[+cor][+nas][+str]
θr/θl	=	[+cont][+cor][+/-retr]	sm	=	[-cor][+nas][+str]
fr/fl	=	[+cont][-cor][+/-retr]	sw	=	[+back][+str]
dr/dw	=	[+cor][+voi][+/-retr]	š	=	[-back][+str]

tr/tw	=	[+cor][-voi][+/-retr]
br/bl	=	[-cor][+voi][+/-retr]
pr/pl	=	[-cor][-voi][+/-retr]
gr/gl	=	[-ant][+voi][+/-retr]
kr/kl	=	[-ant][-voi][+/-retr]

st	=	[+cor][+ant][+str]
sp	=	[-cor][+ant][+str]
sk^w	=	[+rnd][-ant][+str]
sk	=	[-rnd][-ant][+str]

PrC-Core-Ad$_o$ Onsets

spr/spl	=	[-cor][+/-retr][+str]	(Note: The absence
str	=	[+cor][+/-retr][+str]	of stw is the only
skr/skl	=	[-ant][+/-retr][+str]	accidental gap.)

2.5 <u>Summary of syllable theory</u>. The following points summarize the theoretical part of this discussion:

(1) Markedness plays exactly two roles in formal phonology. These are, first, to specify discrete points along parameters, as in (2) and, second, to define the branches in coding trees which are pruned by neutralization rules.

(2) Plus-minus specifications of distinctive features are stored in high nodes of prosodic structure. I have illustrated only syllable structure here, but the point applies to foot structure as well. Words are represented as sequences of feet, where each foot consists of a bundle of distinctive features, with a minimum of order specifications.

(3) The only nodes that are specified in the lexicon are the optional ones.

(4) There is a set of neutralization rules, which have the effect of allowing only the unmarked representatives of the neutralized oppositions to appear in nodes which are specified for certain distinctive features. The features which trigger the neutralization rules are specified in plus-minus terms.

(5) There is an algorithm which generates a candidate set of syllabic templates and traces a unique path from the root of the tree to a terminal element. The shortest template in the candidate set which allows a unique path is selected.

3. APPLICATIONS TO SECOND LANGUAGE ACQUISITION

Many observers have noted that the errors made by adult language learners in attempting to pronounce words of the target language can be characterized as due to the influence of native phonotactic

constraints. Accordingly, we might expect that an adequate theory
of lexical structure should throw light on interlanguage phonology.

3.1 <u>Onsets vs. Codas</u>. It was mentioned above that a given
phoneme may have quite distinct characterizations depending on the
position it occupies in the syllable. Similarly, a consonant
cluster in the onset may have different characterizations from such
a cluster in the coda. Accordingly, if we assume that language
learners may apply native syllabic structure to the words they are
learning, they may respond differently to the same English target
cluster if they come from languages which differ in terms of whether
a coda or an outset analysis is forced upon the cluster.

To test this hypothesis, Cindy Greenberg of Queens College
compared the errors of Greek and Turkish college students learning
English at Queens College. Greek allows syllable initial clusters,
but no final ones; Turkish allows exactly the opposite
configuration. Both groups were presented with tasks forcing them
to pronounce /sp/ sequences in English words in initial and final
positions. Greenberg found that Greeks had no difficulty with
<u>inital</u> /sp/ clusters, but resorted to a variety of strategies to
avoid <u>final</u> /sp/ clusters. The Turks had difficulty with English
<u>initial</u> /sp/ clusters, but <u>final</u> clusters were fine.

3.2 <u>Representations vs. rules</u>. In an interesting paper by Fred
Eckman entitled "Universals, Typologies and Interlanguage," the
important distinction between representations and rules is made,
albeit in a different way from here. In analyzing the speech
production of beginning English learners with a number of native
language backgrounds, he concludes that phonetic sequences in
interlanguage obey universal phonotactic constraints. He finds,
however, that the <u>rules</u> of interlanguage phonology that he proposes
<u>violate purported universals</u>. He explicitly states that the
difference in the behavior of representations (he used the term
<u>forms</u>) and of rules shed some light on the type of system an
interlanguage may be. I agree with this statement, but the
phenomena for which Eckman requires rules are handled in my theory
by universal conventions for spelling out phonemes.

Consider the Japanese learner's rendition of the English word
<u>love</u> in (19). The learner seems to be vacillating between a correct
pronunciation of the word and one with a final schwa. Eckman
handles this by saying that there is an optional rule of schwa
epenthesis in word-final position, and that this rule violates a
putative universal constraint against such rules for all human
languages. For the sake of argument, let's assume there is such a
constraint. Now the questions is, does the Japanese speaker really
apply such a rule?

(19)   Japanese-English Interlanguage Data
(from Eckman, 1984)

IL form	English gloss
lʌv ~ lʌvə	love
dʌn	done
bev ~ bevə	bathe
tUk	took
bæd ~ bæd	bad
lɛt	let
pIg ~ pIgə	pig
tUn	tune
tus	tooth
rud ~ rudə	rude
tʌv ~ tʌvə	tub
lidə	leader
kul	cool

First, let us assume that the variability in pronunciation reflects variability in the learner's ability to pronounce English words correctly; accordingly, an "optional rule" does not seem a logical candidate for an explanatory mechanism. In fact, an epenthesis rule is suggested only if one argues that interlanguage (i.e., in this case the Japanese learner's conception of English) had /lʌv/ as an underlying form and [lʌvə] as the phonetic representation. However, when the Japanese speaker stores the English syllable 'love' in his or her new lexicon for English, it will contain the distinctive features for /l/ and the distinctive features for /v/, along with some specification that the /l/ precedes the /v/. No more information about the consonants need be specified. Let us also assume that the beginning Japanese learner has not yet figured out that English is different from Japanese in the specification of possible syllabic structures. Accordingly, there is only one way of pronouncing this series of consonants. Japanese has the syllable structure rules depicted roughly in (20), along with the coding tree for the coda shown there. The algorithm described above for English yields the form shown in (21). No rule of epenthesis is required, only a universal convention for realizing features stored at high nodes.

(20)   Japanese Syllable Structure Rules
and Coding Tree for Coda

$ ⟶ On + R

R ⟶ V + Cd

Coda
|
[+nas]
|
/n/

(21)   Interlanguage Phonological Representation of <u>love</u>

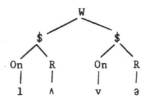

Note that according to this analysis there is no violation of
any universals at all in interlanguage phonology. The only
violations Eckman ever claimed were in the realm of rules, and in my
perspective such rules are not needed to account for the phenomena
at hand.

4. APPLICATION TO FIRST LANGUAGE ACQUISITION

Lisa Menn observes that many (but not all) of the "errors"
children make in pronunciation are due to the severe phonotactic
constraints children obey. In a series of important articles (of
which "Phonological Units in Beginning Speech", 1978, is a good
example), she develops a theoretical model of language processing
and phonological structure which, like Eckman's approach, makes the
important distinction between phonotactic constraints and
phonological rules. Although Menn's approach to phonology is very
different from mine (she adopts a version of autosegmental theory),
many of her insights are key to the development not only of this
application of my theory to first language acquisition, but also to
the development of the theory in general. For example, she cites
the example of a child whose lexicon allows fricatives only in final
position. For this child, the location of /z/ in the lexical
representation of 'nose' is "left unspecified since it is completely
redundant" (Menn, 1978, p. 165). This and other insights were
among the stimuli which led me to formulate a lexical theory which
minimizes reference to the order of elements.

Menn points out that children's speech obeys severe phonotactic constraints which vary from child to child and which are
frequently rather arbitrary. For example, she cites examples of
children who obey consonant harmony constraints so that <u>stuck</u>, <u>duck</u>,
and <u>truck</u> are all pronounced [g k]; these are relative position
constraints which force "the first consonant in a CVC(V) word to be
articulated as far front as or farther front than the second
consonant" (p. 162).

Menn cites many constraints like these; it is evident that a
significant number of the child's phonotactic constraints are

not characteristic of any adult language, although it is safe to say
that all phonotactic constraints which are sufficiently common to
appear among the universals linguists use to calculate markedness
values are evident in child language. It seems, then, that the
child starts off with no marked syllable structure (i.e., at the
"starting point" of parameter structures such as depicted in (3)),
along with an arbitrary set of individual constraints which are
eventually shed. This seems to be a significant difference between
first and second language acquisition, because adult language
learners typically impose phonotactic constraints from their first
language, or from universal linguistic theory, but not arbitrarily.
The child's behavior may be due to the fact that the universal
parameter structure still allows for a greater range of articulatory
distinctiveness than the immature vocal tract is capable of.
Perhaps such speculations can be investigated empirically in the
future, but for now it is well worth noting that there are limits to
analogizing between first and second language acquisition.

Despite these caveats, it is reasonable to suppose that the
child's lexicon is structured in a way basically similar to that of
the adult. Consider Menn's example of the child who obeys the
constraint restricitng fricatives to word final position and who
pronounces the word snow as [nos]. It is reasonable to assume that
this word is represented in the child's lexicon as in (22). The
features [+ round] and [−high] refer, of course, to the vocalic

(22)   Child's Lexical Representation of snow.

$$\begin{bmatrix} \$ \\ + \text{ str} \\ + \text{ nas} \\ + \text{ round} \\ - \text{ high} \end{bmatrix}$$

features. The features of stridency and nasality are not specified
for which segment they belong to, because the latter information is
completely redundant. Assuming that the child has syllable structure rules something like those in (23), along with a constraint
against fricatives in any but final position. The algorithm described in Section 2 of this paper would apply to the representation
(22) of this child's

(23)   Child's Syllable Structure Rules

$$\$ \longrightarrow (On) \quad R$$

$$R \longrightarrow V \ + \ \text{Coda}$$

language to produce the phonological form (24), thus accounting for this child's [nos] as a rendition of the adult snow.

(24)   Child's Phonological Representation of snow

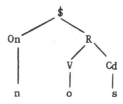

This approach to child phonology and the theory of markedness and prosodic structure on which it is based makes specific claims concerning what children must learn. There are both positive and negative aspects to the child's learning. Positively, the child requires evidence that the ambient language is marked in the sense that it requires marked parameter specifications; thus, the child learning English must hear--in fact, master--an obstruent-liquid onset in order to know that English occupies point (2b) on the upper onset parameter in (3). As the child develops more complex syllabic structures in the lexicon, the early approximations to adult forms, like the child's representation of snow, will end up phonologically more accurate. In many cases, this will require further lexical specifications, because more distinctions become possible as the lexicon grows more complicated. For example, the child's lexical representation of snow will require more specification than indicated in (22) in order to keep this form distinct from a possible (but not occurring) /nos/.

Negatively, the child must learn to lose the early, arbitrary constraints which tend to be idiosyncratic. To venture a speculation, it seems plausible that as the child requires a larger vocabulary, the early severe constraints which are not obeyed by the adult language will be shed in order to provide the possibility for more distinctions. A valuable area of future research is the study of the loss of these early constraints in child phonotactics.

5. CONCLUSION

This paper presents a theory of lexical representation which takes seriously Menn's suggestion that ordering redundancies are as significant an aspect of phonotactics as are constraints on classes of sequences which may appear in a given position. Notions of markedness and prosodic structure have been incorporated into an overall theory which attempts a concise account of a variety of phonotactic phenomena.

In the past few years, researchers in first and second language acquisition have devoted serious attention to the study of phonotactic constraints. Both Eckman and Menn have emphasized the importance of distinguishing between phonotactic constraints and phonological rules as aspects of the language learners' output. Eckman in particular has claimed that although interlanguage forms obey universal aspects of phonotactics, interlanguage rules may be deviant (but not arbitrarily so). The theory advocated here is consistent with the claim that not even the rules of interlanguage disobey universal principles, because Eckman's examples can all be accounted for by the algorithm for converting lexical into phonological representations. No phonological rules need be invoked here, nor in first language acquisition either.

Among the main goals of this paper has been that of attempting an integration of a theory of synchronic phonology with first and second language acquisition. It has often been said that markedness theory should shed light on language acquisition. As the theory has evolved over the past several years, it has become possible to suggest more constrained hypotheses not only of synchronic structure, but also of the application of such a hypothesis, in that it provides a tightly constrained theory of lexical and phonological structure, as well as a natural application to language acquisition.

REFERENCES

Cairns, C. 1971 Review of Principles of Phonology (by N. S. Trubetzkoy), translated by Christine Baltaxe, Language, 47, 918-931.

Cairns, C. 1977   Universal Properties of Umlaut and Vowel Coalescence Rules: Implications for Rotuman Phonology.   In Alphonse Juilland (ed) Linguistic Studies Offered to Joseph Greenberg.   Saratoga, Calif: Anma Libri.

Cairns, C. 1982 "Prosodic Structures in the Lexicon," paper presented at the Conference on Distinctive Features, State University of New York at Stony Brook, May 1981.

Cairns, C. and M. Feinstein 1982 "Markedness and the Theory of Syllable Structure," Linguistic Inquiry, 13, 193-225.

Eckman, F. 1984 "Universals, Typologies and Interlanguage," In W. E. Rutherford (ed.) Language Universals and Second Language Acquisition, 79-105.   Amsterdam: John Benjamins Publishing Company.

Ferguson, C. 1963 "Assumptions about Nasals: A Sample Study in Phonological Universals," in J. Greenberg (ed.) Universals of Language, Cambridge: MIT Press, 42-47.

Greenberg, C. 1983 "Syllable Structure in Second Language Acquisition," in press in CUNYFORUM, City University of New York: New York.

Greenberg, J. 1960 Universals of Language. The Hague: Mouton.

Greenberg, J. 1978 "Some Generalizations Concerning Initial and
    Final Consonant Clusters," in J. Greenberg, C. Ferguson, and E.
    Moravcsik (eds.) Universals of Human Language, vol. 2:
    Phonology, 243-279. (Originally published in 1964 in Russian in
    Voprosy Jazykoznanija and subsequently in English in 1965 in
    Linguistics.)
Jakobson, R. 1939 "Signe zéro," Mélanges Bally, 143-152.
Kean, M. L. 1975 The Theory of Markedness in Generative
    Grammar, Doctoral dissertation, MIT, Cambridge, Massachusetts.
Menn, L. 1978 "Phonological Units in Beginning Speech," in A. Bell
and J. Hooper (eds.) Syllables and Segments, Amsterdam: North-
    Holland, 157-171.
Rutherford, W. 1982 "Markedness in Second Language
    Acquisition," Language Learning 32, 85-108.

MARKED VOWEL SYSTEMS AND DISTINCTIVE PALATALIZATION

Thomas J. Cox

San Diego State University

While the term 'palatalization' is used in many different ways in the pertinent literature, this study is limited to consideration of those processes which convert plain obstruents to palatalized [c] = [ts] or [ɜ] = [dz] as opposed to those processes which result in palatal consonants such as [č], [ǰ], [š], or [ž]. Its central claim is that the circumstances for such developments are to be found in the marked nature of the vowel systems of palatalizing languages and in the disruption of redundancy conditions with such markedness involves.

In the standard theory elaborated by Chomsky and Halle (1968) the palatalization of obstruents involves assimilation of plain consonants to following front or high segments. According to this formulation, palatalized [c] and [ɜ], which are not ordinarily classified as high segments, must be so designated just where they occur as the output of a palatalization rule. However, if palatalization is seen instead as an accommodation in manner of articulation of the major class of consonants to the major class of vowels, and alternative analysis is possible.

It is a universal phonetic circumstance that consonant + vowel sequences are characterized by the presence of nearly imperceptible vocalic transitions which can be noted thus: $[t^i i]$. In light of another linguistic universal, i.e. the preference for consonant clusters over vowel clusters, the shift from sequences such as $[t^i i]$ to $[t^s i]$ is best understood as a shift of a minimally specified vocalic transition to a minimally specified consonantal one. The relative height of such vocalic transitions is not distinctive; for example, the transition element in the sequence $[k^e e]$ is not a high segment but, should certain circumstances convert that element to a glide,

that glide would automatically be marked [+high] in light of its
function. Although these vocalic transitions occur in all languages, they are not usually distinctive in function, nor do
they usually bring about palatalization of preceding obstruents;
most languages simply retain plain consonants in such environments or show fronted velars. This suggests that the conditions
for distinctive palatalization, rather than being narrowly phonetic, are in fact structural in nature. Comparisons of two
unrelated languages with similar vowel systems -- Japanese,
which is a palatalizing language, and Spanish, which is not --
provides some insight into the nature of those conditions.

Spanish has four non-low vowels: /i,e,o,u/, and a single
low vowel: /a/. Since rounding is a redundant feature of
non-low back vowels, Spanish has, from a phonological
perspective, only plain consonants, phonetically palatalized
before front vowels and phonetically rounded before non-low back
vowels.

In Japanese, on the other hand, /t/ and /d/ palatalize to
[c] and [3] before /u/ and to [$t^{s'}$] and [$d^{z'}$] before /i/.
The fact that palatalization of dentals occurs before /u/ as
well as before /i/ in that language draws our attention to an
important contrast with Spanish; Japanese /u/ is a high, back,
unrounded vowel, a circumstance usually described as phonetic,
but one which nonetheless has phonological implications for that
language.

According to the historical account provided by Miller
(1967), by the beginning of the tenth century, Japanese had
moved from and eight-vowel system to what is essentially the
modern five-vowel one. From the fact that the formation of
palatal /š/ from /s/ took place before Old Japanese /e/ and /i/
but not before /u/, it may be deduced that /u/ was still a round
vowel at that stage and that the usual redundancy conditions applied. The unrounding of /u/ would have disrupted these conditions and, in fact, Middle Japanese shows palatalized dentals
before both /i/ and /u/. Apparently, when /u/ unrounded, the
link between phonetic palatalization and the occurrence of front
vowels disappeared; what had previously been a strictly phonetic
contrast between [t] and [t^i] became a phonological one and
further development [t^s] represents the unmarked result of
such phonologization. Although the unrounding of /u/ is admit-
tedly a linguistic rarity, disruptions of back vs. non-back redundancy conditions are not, and further data from other palatalizing languages point to the crucial role of such developments.

In order to evaluate that data, however, the notion of
markedness as applied to vowel systems must first be considered. While languages appear to show a preference for non-

complex segments, the relative frequency of five-vowel 'cardinal'
systems of three-and seven-vowel triangular systems suggests
that systematic markedness is more closely related to the degree
of asymmetry than to the marked nature of individual segments.
However, while the standard theory recognizes the need for including symmetry conditions in the marking conventions for vowel
systems (Chomsky and Halle: 410), it also suggests that symmetry
may be considered a feature of even-numbered quadrangular vowel
systems. But, in contrast to three-, five-, or seven-vowel systems, quadrangular vowel systems are only symmetrical in respect
to some imaginary point, whereas the point of reference in triangular systems is always one of the segments, i.e., the maximally
unmarked vowel /a/. Although it is true that quadrangular vowel
systems have an equal number of back and non-back vowels, the
primary distinctive opposition is such systems is usually round
vs. non-round and since /a/ is at best neutral in this regard,
quadrangular systems are inherently asymmetrical. (Trubetzkoy
1969:100)

Therefore it must be specified that the symmetry typical of
unmarked vowel systems is of a particular kind: it involves triangularity and opposition based on the feature [± back]. For example, while languages like Modern French and German show relatively marked, front-rounded vowels, their vowel systems are in
fact relatively unmarked; they are symmetrical in respect to the
low vowel /a/ and reflect a basic opposition based on the back
vs. non-back distinction. As our hypothesis predicts, neither
French nor German is a palatalizing language and data from other
languages and language groups provide similar confirmation.

Triangular vowel systems are so extensively represented in
the large family of Bantu languages that such symmetry is
considered an important classificatory criterion for the
identification of these languages. Significantly, although
palatals are common in Bantu languages, a recent extensive survey
of African languages (Welmers 1973) does not show a single Bantu
language having palatalized consonants. That study does,
however, classify Tiv, a palatalizing language as belonging to
the Benue-Congo group of the basis of its quadrangular vowel
system whose primary opposition is round vs. non-round.

Within the family of Germanic languages, the link between
shifting redundancy rules and distinctive velar fronting points
to the way in which distinctive palatalization arises. Whereas
German, with its symmetrical vowel system, reflects the common
pattern of redundant velar fronting, Modern Icelandic shows an
unusual pattern of distinctive velar fronting before /a,o, and u/,
plain velars before phonetically fronted [ø] and [y] and fronted
velars before the diphthong /ae:/. These developments can best
be understood in light of shifting reduncancy conditions over

the long history of Icelandic. According to Haugen (1976) in
Old Norse, original long /u/ first fronted to /y/, later rounded
to /i/ and eventually merged with original long /i/ in Modern
Icelandic.

The fronting of long /u/ to /y/ must have disrupted a
symmetrical vowel system whose primary opposition was back vs.
non-back since the velar in the Old Norse [kynna] *KUNNAN 'to
come to know' for example, was not fronted. At that point,
velar fronting was predictable on the basis of rounding rather
than fronting. Although the eventual merger of /y/ and long /i/
re-established back vs. non-back redundancy conditions, the
range of environments in which velar fronting then took place
made its occurrence impossible to predict on the basis of either
opposition. The result was that Icelandic pronunciation began
to be characterized by a non-conditioned fronting of velars
except before original round vowels, where the contrast between
fronted and plain velars continued to have a distinctive
function. All subsequent borrowings containing velars and
velars before the new diphthong [ae:] were subject to the
general velar fronting rule. As for Modern Icelandic [y] and
[∅], these are actually phonetically fronted versions of /u/
and /o/ and represent the structural equivalent of velar fronting; the front vowel is a substitution for velar fronting in an
environment where such fronting is blocked. Anderson (1981)
points out that the fronting of these vowels took place at a
relatively recent period; in fact it must ·have taken place at a
stage when the primary opposition was still round vs. non-round,
that is, during the period when /y/ was still distinct from long
/i/.

The predictable result of the later merger of /y/ with long
/i/ in Icelandic is that distinctive palatalization is not a
feature of that language. Once **kynna** became **kina** [k'ina] there
were no longer any instances of back velar + front vowel
sequences (except as the result of phonetic fronting) and the
back versus non-back opposition was restored.

As an example of the contrasting effects of redundancy
conditions on distinctive palatalization within another language
family, we can contrast Modern Russian with Modern Czech. In
the palatalizing dialects of Russian we find marked vowel
systems which include front rounded /y/, whereas in
on-palatalizing Czech, the symmetrical system of long and short
vowels preserves the balance between phonetic palatalization
before front vowels and phonetic rounding before round vowels.

The phonetically fronted variants of Russian vowels which
occur after palatalized consonants reveal that the back vs.
non-back opposition does not play the primary distinctive role

in the Russian system.   In Czech, however, with its triangular
system of long and short vowels the back vs. non-back is
preserved and palatalization is entirely predictable in terms of
the feature.   The effects of these differences can be seen in
the Russian and Czech substitutes for loan words from French or
German having front rounded /y/ or /ö/.   Russian substitutes /u/
and /o/ respectively but palatalizes the preceding consonant,
the palatalization alone signaling the fronted variants of the
rounded vowels; Czech substitutes /i/ and /e/ respectively, the
front vowel signaling palatalization.   Thus French <u>bureau</u>
'office' is realized as [b'uro] in Russian and (biro:) in
Czech.   In such environments, Russian palatalization represents
the structural equivalent of vowel fronting in Icelandic.

   Shifting redundancy rules may also explain the diachronic
split between palatalizing Western Romance and non-palatalizing
South and Eastern Romance.   Classical Latin, with its
symmetrical system of short and long vowels, was not a
palatalizing language.   If it were not for the fact that velar
palatalization occurred in Ibero-Romance as well as in
Gallo-Romance and the Northern Italian and Rhaeto-Romance
dialects, the shift of Latin /u/ to /y/ which took place in the
latter might be held to account for those developments.
However, by all accounts the Gallo-Romance shift from /u/ to /y/
was not completed until after Latin had begun to separate into
dialect area, even though velar palatalization before /i/ and
/e/ is a shared development in Western Romance.

   Although the phonetic and phonological facts of Vulgar Latin
are not entirely know, /y/ must have co-existed with Latin /u/
over a long period of time, if only as one pronunciation of the
many Greek loanwords such as GYRUS which, according to Grandgent
(1970:80), was correctly pronounced [gyrus] by 'cultivated
people'.   The very presence of such a vowel and its effect on
redundancy rules could well have been the destabilizing facor
which wrought such havoc with the Classical Latin vowel system;
its presence would have led to confusion among /i/, /y/ and /ĭ/
when length ceased to be distinctive.   As a result, the
distinction between [i] and [ĭ] was re-interpreted in terms of
relative height, bringing about the eventual merger of short /ĭ/
and unstressed long /e/.   It is difficult to imagine, in the
absence of some such structural asymmetry or imablance, what
would have blocked the more reasonable merger of long and short
/i/ such as that which occurred in more conservative dialects
such as Central Sardinian.

   In any event, it is unnecessary to assume that /u/ fronted
to /y/ in all of Western Romance to make the case that there was
such a vowel in Vulgar Latin and that the resultant
destabilization of the relevant redundancy rules produced the

prerequisite conditions for distinctive palatalization. Again, we have only to compare conservative Sardinian, with its cardinal vowel system in which no back-to-front or high-to-mid vowel changes occurred and where there are no velar palatalizations, to infer a crucil role for vocalic asymmetry in the rise of velar palatalization in Western Romance.

A synchronic example of the disruptive effects of the breakdown of redundancy conditions may be observed in Québecois as described by Gendron (1966). In contrast to non-palatalizing Standard French, Québecois is characterized by a vocalic system whose redundancy rules are in the throes of readjustment. The fact that phonetic palatalization of velars occurs everywhere except before /u/ and /o/, suggests a primary opposition based on rounding, but since dental palatalization occurs before both /y/ and /i/, that rounding does not appear to block palatalization. But in fact, the vowel system of Québecois is in the process of becoming asymmetrical; both /i/ and /y/ are subject to devoicing and eventual syncope following a palatalized consonant as reflected in the pronunciation [parcipe] for __participer__ 'to participate'; the distinction between /i/ and /y/ is effectively neutralized in that environment and, in other environments, /i/ is shortening (or lowering) to /I/, resulting in the pronounciation [eglIz] for French __église__ 'church'. Even though such changes are far from universal in Québecois, they are widespread and suggest a state of affairs in that language similar to the one which must have existed during the early palatalizing period in Vulgar Latin.

While some of the data presented here might equally well be interpreted to suggest that palatalization of obstruents brings about vowel fronting or other vocalic asymmetries, consideration of the developments in Japanese where /u/ is not fronted, but rather unrounded, argues against reversing the order of cause and effect. Furthermore, it must be recalled that in Icelandic, where both /u/ and /o/ are phonetically fronted to [y] and [ɸ], respectively, distinctive palatalization (as opposed to distinctive velar fronting) does not occur.

What this study claims, then, is that the definition of an unmarked vowel system must be linked to conditions which promote the stability of consonants. More specifically, it claims that a particular kind of symmetry involving triangularity and a primary opposition based on the feature [+back] vs. [-back] characterizes the optimally unmarked vowel system and that disruption of this opposition is prerequisite to the distinctive palatalization of obstruents. Where further data can be adduced to the contrary, some refinement may be necessary, but the substantial agreement to be seen among the major palatalizing languages and language groups is presented here in support of the basic validity of these claims.

REFERENCES

Anderson, Stephan. 1981. Why Phonology Isn't Natural.
    Linguistic Inquiry 12. 493-539.

Chomsky, Noam and Morris Halle. 1969. The Sound Pattern of
    English. New York. Harper & Row.

Gendron, Jean. 1966. Tendences phonétiques du français parlé au
    Canada. Québec. Les Presses de l'Université Laval.

Grandgent, Charles. 1907. An Introduction to Vulgar Latin.
    Boston. D.C. Heath.

Haugen, Einar. 1976. The Scandinavian Languages. Cambridge.
    Harvard University Press.

Miller, Roy. 1967. The Japanese Language. Chicago. University
    of Chicago Press.

Trubetzkoy, Nikolai. 1969. Principles of Phonology. trans. by
    Christiane Baltaxe. Berkeley. University of California
    Press.

EPENTHESIS AND MARKEDNESS

Karen Scriven

The University of Iowa

0. This paper develops an analysis of epenthesis as a universal syllable-based process in which syllable boundaries arise through the application of markedness constraints on consonant sequences at the level of lexical representation. The purpose of the analysis is to show that, universally, epenthesis occurs in order to create nuclear material for "stranded" consonants, that is, for consonants which are separated from any vocalic material because of constraints on the placement of syllable boundaries. The analysis also explains where epenthesis occurs: either to the left or the right of a syllabically isolated consonant, where direction is dependent on the existence or nonexistence of a language specific constraint on resyllabification.

1. The capacity to refer to syllable structure not only simplifies the description of many phonological rules, but allows such rules to be described for the first time in ways that capture meaningful generalizations. In Spanish (Harris, 1969), for example, nasals assimilate before obstruents in the same word or across word boundaries:

(1)   un beso   [umbeso]   'a kiss'
              un charco   [uñčarko]   'a pool'
              un gato   [uŋgato]   'a cat'

However, before glides, nasal assimilation occurs only across word boundaries:

(2)   miel   [myel]   'honey'
              nyeto   [nyeto]   'grandson'
              un hielo   [uñyelo]   'an ice'
              un huevo   [uŋweβo]   'an egg'

Hooper (1972) notes that the obligatory presence of the word boundary
in the second rule means that the two nasal assimilation rules can
not be collapsed, even though they obviously describe the same phenomenon. However, if the syllable boundary is included in the
string, it becomes apparent that nasal assimilation occurs before a
consonant or a glide only if a syllable boundary intervenes. A
natural rule can be written which blocks nasal assimilation in the
environment where the nasal and following glide are tautosyllabic.

   In addition to the rule simplification, an· important argument
for the need to refer to syllable structure in phonological descrip-
tion is based on the abbreviation $\left\{ \begin{matrix} C \\ \# \end{matrix} \right\}$ (Kahn, 1976). While this is
a frequently used abbreviation, the two elements in the braces have
no shared features. But more importantly, every rule which can be
written using $\left\{ \begin{matrix} C \\ \# \end{matrix} \right\}$ can more accurately be written by referring to
syllable boundaries. Hooper (1972) discusses the nasalization of
high vowels in Akan (Schacter & Fromkin, 1968):

(3) $V \longrightarrow [+ \text{nasal}]/$ _____ $[+ \text{nasal}] \left\{ \begin{matrix} C \\ \# \end{matrix} \right\}$

        $[+ \text{hi}]$

noting that it is not only simpler to replace the $\left\{ \begin{matrix} C \\ \# \end{matrix} \right\}$ by a syllable
boundary, but that the rule with a syllable boundary in its struc-
tural description is the real statement of the environment, while
the rule with $\left\{ \begin{matrix} C \\ \# \end{matrix} \right\}$ is "merely an ad hoc contrivance that produces
the same results" (Hooper, 1972:533).

   It is important to note that syllable structure is used as a
basis for the operation of phonological rules, and therefore must be
assigned prior to the application of such rules. But how is syllable
structure to be assigned?

2.   Several proposals have been made to predict the permissable
sequence of segments inside syllables, and thus the delimitation
between syllables. While none of these proposals is adequate as a
universal predictor of syllable boundaries, one of these proposals
does provide a largely universal prediction of syllabic onsets.
Cairns and Feinstein (1982) abandon the attempt to find a single
scale of markedness to account for all syllabic onset clusters.
Rather, they point out that though single consonant onsets are
maximally unmarked, there is no evidence to suggest that onset
clusters can be analyzed along a single scale (Greenberg, 1966).
There are, in fact, three paths a language may follow in complicating its syllabic inventory:

(4)

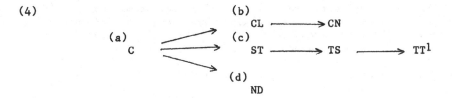

```
C = obstruent
L = liquid
N = nasal
S = fricative
T = stop
D = voiced
```

(a) is the maximally unmarked syllabic onset. (b), (c), and (d) are all permissable onsets, yet share no markedness relationship (Greenberg, 1966); rather, a language may follow any one of these paths in complicating its syllabic inventory. Furthermore, a language may take more than one path, as in modern English which has both CL and ST clusters. In addition, languages appear to allow "hybrid" complex onset types, like the substitution of ST in CL onsets in English, or the substitution of ND for C in CL onsets in languages with both nasal-stop clusters and obstruent-liquid clusters.

While the Cairns and Feinstein proposal is adequate as an analysis of word-initial syllabic onsets, it is only partially adequate for delimiting word-internal syllable boundaries. To predict word-internal syllable boundaries, we must turn to a consideration of Lowenstamm's (1981) "typical syllable" which ranks segments according to the position they occupy within the syllable, going from the nucleus to the margins, and from the most sonorant to the least sonorant segments:

(5)   vowel
         glide
         sonorant
         fricative
         stop

The typical syllable is subject to two universal restrictions which will be adopted here. First, no segment in a syllable is lower on the hierarchy than both of its immediate neighbors. Thus, a sequence /rdr/ could never be tautosyllabic, since /d/ is lower than both of its immediate neighbors. This restriction may be modified to allow for NDL and STL clusters by adopting Cairns and Feinstein's assumption that in such cases ST and ND merely represent a "complex" C.

The second universal restriction which Lowenstamm proposes is
that no two segments of equal ranking may be adjacent inside a
syllable. Thus, in French we can predict the syllable boundaries
whenever two adjacent segments are of the same ranking:[2]

(6)   chaos   [ka.o]   'chaos'

               actif   [ak. tif]   'active'

               asphalte   [as. falt]   'asphalt'

Lowenstamm's two universal restrictions will be invoked frequently
in determining impermissable consonant cluster sequences.

This analysis assumes that there is a language specific choice
involved in whether onset inventories are complex or not, and in
whether these are constructed according to path (b), (c), or (d).
Thus, for any language, the grammar will contain a statement as to
whether the language is of type (a), (b), (c), or (d), then syllable
boundaries are established according to these categories and in line
with Lowenstamm's two universal restrictions. Additionally, it is
assumed that onsets are maximized to the degree allowed in the
language in question (Kahn, 1976). That is, first each vowel is
designated as a syllabic nuclei, then as many [-syll] segments as
create permissable initial clusters form the onset of any particular syllable. In this way, syllable boundaries arise when a sequence
of consonants violates the restrictions on consonant clusters in
that language type. In assigning syllable structure to contradiction, the process would be as follows:

(7)

Notice that /n/ and /k/ are not assigned as onsets since there are
no stop-fricative clusters, nor nasal-stop clusters in English.
/n/ and /k/ are thus assigned to the coda of the preceding syllable.
Most of the research in syllable structure has been concerned with
the nature of syllabic onsets and there is little discussion of the
structure of or constraints on the coda. It will be assumed in this
analysis (Greenberg, 1966) that the least marked syllable is open,
that is, ends in a vowel. In languages with complex types it is
assumed that permissable word internal codas are the mirror image of
permissable onsets for each of the three language types.

3. If syllable boundaries arise when sequences of consonants violate
the markedness restriction on consonant clusters, it follows that in
lexical representation consonants may be "isolated" in syllables
with no nuclear material. However, in surface structure, all
syllables must have a syllabic peak (nuclear material). Epenthesis

can be seen to apply at the site of a syllable boundary when a consonant has been isolated by the operation of syllable assignment
processes and is not associated with any nuclear material.

Thus, in Hebrew (Lowenstamm, 1981):

(8)   g. dɔ. rəm.   'limitations'

syllable assignment processes, based on markedness constraints, have
isolated [g] into a syllable with no syllablic peak (nucleus).
Epenthesis applies to supply that peak, creating the surface form.

The term epenthesis can, under this analysis, be used to describe
both the insertion of vowels and the assignment of syllabic values
to consonants, since in both cases a syllable is created out of a
consonant "stranded" by the operations of markedness constraints.[3]

Let us examine several instances of epenthesis to verify that
epenthesis does always occur at a syllable boundary. In English,
the syllable breaks in a word like ladder are [læ d . r]. [dr] is
not an acceptable codas, since [rd] is not an acceptable onset in
English. This analysis clarifies why resonants become syllabic in
English: they are "stranded" consonants without the nuclear
material which every syllable must have at the surface level. This
analysis also explains why the syllabic [1] in a word like simple
does not remain syllabic in simply, since there is now nuclear
material in the syllable containing [1].

In Yawelmani (Kenstowicz & Kisseberth, 1977), there is a well
known vowel epenthesis rule:

(9) $\emptyset \rightarrow i \: / \: C\_\_\_\_C \begin{Bmatrix} C \\ \# \end{Bmatrix}$

This rule applies to underlying /:a:mlhin/ ' help' to produce
/:a:milhin/ and to underlying /xatt/ 'eat' to produce /xatit/. One
can recognize in the structural description of the rule that environment which was originally proposed as evidence for the need for syllables in phonological theory. If we assume a lexical representation which includes the syllable boundary markings:

(10)   /?a:m. 1. hin/

          /xat. t/

We can see that epenthesis does in fact occur at the boundaries of
syllables. Notice that these syllable boundaries once again follow
the universal syllables that we have established, since any other
syllabification would violate Lowenstamm's second univeral restriction that no two segments of equal ranking may be adjacent inside a
syllable.

All cases of epenthesis of which I am aware can be analyzed in terms of the operation of a single process serving to provide nuclear material for marked consonant cluster sequences which can not be tautosyllabic. This contradicts Broselow (1983), who divides epenthesis into separate processes. For Broselow, syllabically conditioned epenthesis applies when a string of segments in under-lying form does not constitute a sequence permissable in a well-formed syllable. Segmentally conditioned epenthesis, on the other hand, occurs when a particular sequence of segments is disallowed and functions simply to maintain phonotactic constraints in the language[4].

To illustrate the difference, in Swahili syllabically condi-tioned epenthesis occurs in borrowed words to insure that all syllables are of CV structure:

(11)   tiket ——→ tiketi   'ticket'

            ratli ——→ ratili   'pound'

This can be contrasted with Winnebago, where segmentally conditioned epenthesis serves to break up obstruent-sonorant clusters:

(12)   hošwaža   hošawaža   'be sick, 2nd pers'

However, if the analysis presented here is assumed to apply to these forms, the distinction between the two processes disappears, since in both cases epenthesis occurs to create nuclear material for stranded consonants. Thus, in Swahili:

(13)   ti. ke. t ——→ ti. ke. ti

            ra. t. li. ——→ ra. ti. li

Winnebago, on the other hand, does not permit tautosyllabic obstruent-sonorant clusters. Thus, in Winnebago as in Swahili we are dealing with a violation of what can constitute a well-formed syllable in the language:

(14)   ho. š. wa. ža.   ho. ša. wa. ža[5]

Broselow bases the distinction between types of epenthesis on the interaction between stress assignment and vowel insertion. However, as Broselow herself points out, this predicted interaction still con-tains exceptions. Segmentally conditioned epenthesis may either precede or follow stress assignment, depending on the language. Syllabically conditioned epenthesis usually precedes stress assign-ment, but the ordering may be reversed depending, according to Broselow, on the "internal structure of syllables." Broselow 's distinction between types of epenthesis does not solve the problem

of the interaction between stress assignment and epenthesis.
Furthermore, the assumption of a syllable based analysis such as the
one proposed here undermines the distinction between the two epenthesis processes.

4. Having established that syllable boundaries may be determined by
the separation of highly marked consonant sequences, and that epenthesis always occurs at such boundaries, we must now address the
issue of how the relationship between epenthesis, syllable structure, and markedness can best be captured by phonological theory.

Early work on syllable structure (Kahn, 1976) assumed that every
syllable must contain a nucleus. However, such an assumption leaves
material "unsyllabified" in many languages, and leads to a paradox
in the assignment of syllable structure and the operation of epenthesis (Lowenstamm, 1981). By Kahn's analysis, in the assignment of
syllables to the Hebrew gedarem, 'limitations', syllables are first
assigned as follows:

(15)   *   σ   σ
          | /| /|\
          g d ɔ r ə m

/g/ can not be assigned to a separate syllable, since consonants are
assigned only to syllables containing nuclei. A nucleus can be
created by epenthesis only in the environment of a syllable boundary.
But since there can be no syllable boundary, there can be no application of epenthesis. Thus, the correct syllable boundaries can not
be assigned. This ordering paradox arises time and again in the discussion of how to deal with "unsyllabified" consonants.

If underlying syllables are based on the separation between
marked consonant sequences regardless of the existence of nuclei,
then units definable as syllables reemerge as boundaries between
these impermissable cluster sequences. In (15), a syllable boundary
would exist in the lexical representation between /g/ and /d/, since
Hebrew allows TS but not TT onsets. The ordering paradox inherent
in Kahn's work disappears if syllables are not dependent on nuclear
material but arise instead from the separation of impermissable cluster sequences into different syllables.

It has previously been proposed that this ordering paradox could
also be eliminated by the creation of null segments.[6] Kaye and
Lowenstamm (1979) assume that all syllables consist of an onset and
a rime, but that in the lexical entry the rime of certain syllables
may be null.[7] If this is the case, according to Kaye and
Lowenstamm, epenthesis will apply and the rime will be realized as a
schwa. This accounts satisfactorily for the Hebrew cases with which
they are immediately concerned. For example, the underlying repre-

sentation of [xətsufəm] 'impudent person' is:

(16)

$$\sigma \quad \sigma \quad \sigma$$
x ϕ   ts u   fəm

Kaye and Lowenstamm then formulate epenthesis as a syllable sensitive rule which converts the null rime into a schwa.

But Kaye and Lowenstamm do not address the issue of languages in
which epenthesis occurs to the left of a consonant (or syllable)
rather than to the right. For example, in Iraqi Arabic (Broselow,
1983), epenthesis is always to the left:

(17)   katab + t + la ⟶ katabitla   'I wrote to
                                                           him'

Thus, Kaye and Lowenstamm's analysis that all syllables have an
onset but may lack a rime is erroneous. In fact, there are many
cases where epenthesis occurs to the left rather than the right of
an unsyllabified consonant. One of the best known of these is
Spanish, where [e] is inserted before initial s-stop clusters.
Thus, in Hooper's analysis (1968:538), underlying /slabo/ 'Slav' and
/sbelto/ 'svelte' undergo e-epenthesis to produce the surface forms
/es.la.bo/ and /es.bel.to/.

Kaye and Lowenstamm say that epenthesis occurs when a consonant is
in a syllable whose rime (vocalic element) is null. But as we have
seen, such an analysis can not be restricted so that the consonant
in question is always the onset and never the coda, since many languages have epenthesis to the left rather than the right. If epenthesis is instead simply a result of the division between marked
consonant sequences, we can eliminate the introduction of additional
theoretical machinery such as the highly abstract null segment.
Anderson (1982), in a proposal similar to that of Kaye and Lowenstamm, argues that the null segment is necessary in French to adequately distinguish the schwas which alternate with zero (unstable
schwa). Thus, the word pelouse, 'lawn', shows a vowel [ə][8] between
its first two consonants in some environments (cette pelouse [sɛtpə
luz], 'this lawn'). By contrast, a word like seulette 'lonely,
fem.', shows a vowel [ə] in all environments ([səlɛt]).

According to Anderson, the schwas which sometimes delete are o
underlyingly, while those which always surface are underlying [œ].
Anderson then attempts to predict when ø will be deleted rather than
realized as [ə]. ø is deleted when a syllabification rule operates
to strand ø as the only item in a syllable. According to Anderson,

petite, 'small' is underlyingly:

(18)

$$
\begin{array}{ccc}
\sigma & \sigma & \sigma \\
/\backslash & /\backslash & /\backslash \\
p\ \emptyset & t\ i & t\ \emptyset
\end{array}
$$

a resyllabification rule operates to close the middle syllables:

(19)

$$
\begin{array}{ccc}
\sigma & \sigma & \sigma \\
/ & /|\backslash & | \\
p\ \emptyset & t\ i\ t & \emptyset
\end{array}
$$

The final $\emptyset$ is then deleted since it is the only item in the syllable. The first $\emptyset$ is realized as [ə], since every syllable which has
a consonant must also have a nucleus.

Anderson's justification of the null segment is based in part on
Dell (1973) who argued that unstable schwas can not be inserted by
epenthesis since in a word like pelouse the proposed epenthesis rule
would insert [ə] between word initial [p] and a following [l]
(giving, e.g. cette pelouse, [sɛtpəluz]; however, in other words
(e.g. place 'square') epenthesis must be prevented in the same
environment (cette place, 'this square' is always [sɛtplas] never
[s tpəlas]). However, forms like these represent no serious problem
if epenthesis is seen as an operation governed by syllable boundaries. In any language there will be constraints governing possible
consonant cluster sequences. However, in any language which allows
consonant clusters, those clusters will sometimes nonetheless be
broken up by a vowel, and that vowel will sometimes be identical to
the epenthetic (or maximally unmarked) vowel of the language. Dell's
argument is similar to saying there is no epenthesis in English since
no epenthesis rule could explain the contrast between English bout,
about (careful pronunciation), and 'bout (casual pronunciation).
But there is certainly no necessity for positing a null segment as
opposed to epenthesis solely on the basis of of these cases. What
of Anderson's other justifications for distinguishing [ə] from $\emptyset$?

Anderson uses $\emptyset$ in part to prevent schwa deletion before h-
aspire. But as Iverson (1982) points out, schwa sometimes occurs
before h-aspire. Thus, Anderson's rules (together with a rule of
optional schwa-insertion) will interact in the same way whether the
lexical representation contains schwa or $\emptyset$. This considerably weak-
ens the motivation for null which Anderson presents, especially when
Anderson's $\emptyset$ necessitates considerable additional machinery in terms
of resyllabification processes.

In fact, the underlying null which Anderson posits can be eliminated once it is recognized that syllables do in fact often contain

no nuclear material at the underlying level. Thus, those cases where
o deletes in Anderson's terms can more clearly be seen as cases where
the consonantal material in question is already syllabified with nuclear material. For example, in petite [p.ətit], there is no need to
insert a final schwa. That such schwa-insertion is often optional,
especially when consonant clusters are concerned, provides interes-
ing support for the analysis which sees epenthesis as a process
breaking up highly marked sequences of consonants. Thus, according
to Tranel (1981:289), des contacts penibles, 'painful contact' is
often pronounced [də kɔ̃taktəpenibl]. In this case, there is no
morphological motivation for the schwa (as there is in many of the
instances cited by Anderson). Such schwa insertion becomes clearly
related to a perceived difficulty of articulation which is eased by
schwa-insertion in between syllable boundaries.

What about cases where Anderson does insert schwa? In achete,
'he buys', the syllable boundaries are /aš.ət/ and schwa is inserted
before the word final consonant. Note that this analysis is simpler
than Anderson's. In this analysis, syllable boundaries are assigned
by universal markedness constraints and are available as epenthesis
sites; we don't have to go through repeated processes of resyllabification and deletion to arrive at the surface phonetic form.[9]

Thus, if impermissable cluster sequences are assigned to syllables regardless of the existence of nuclear material, we can explain the epenthesis process without creating an ordering paradox,
and without necessitating additional and ad hoc theoretical machinery such as the null segment.

5. There are some apparent problems with this analysis of epenthesis
and its dependence on syllable structure which need to be dealt with.
The restriction of epenthesis to the site of syllable boundaries
might seem superficially similar to the approach of Selkirk (1981),
but there are important differences between the two analyses.
Selkirk assumes that epenthesis serves to provide a vocalic nucleus
for consonants which can not otherwise be incorporated into existing
syllables. But it will be recalled that such an analysis contains a
paradox, since syllables are created on the basis of how the word is
syllabified. Additionally, Selkirk, who maintains that epenthesis
occurs obligatory to break up impermissable clusters, can not
explain optional epenthesis in words like wrestling which may be
prounounced either [rɛsəlIŋ] or [rɛslIŋ]. According to Selkirk,
epenthesis only occurs to break up impermissable clusters, but in
this case, the cluster broken is perfectly acceptable in the
language.

If syllables are assigned at the level of lexical representation
and syllable boundaries are available as epenthesis sites, then
there is nothing in principle to prevent a situation in which epenthesis is an optional process as in French optional epenthesis in

phrases like <u>envie de te le demandier</u> [ɔ̃vi#də#tə#lə#dəmɔ̃die] 'desire
to ask you it', which has eight possible pronunciation variants
(Anderson, 1982), since any of the schwas may be present or not so
long as the constraint against tautosyllabic triconsonantal clusters
is not violated. This case would also be problematic for Selkirk,
since epenthesis can apply or not at any individual site, and the
only constraint is on the number of epenthesis sites which can be
left unfilled.

    A second problem with the analysis proposed here is the apparent
overgeneration of epenthesis sites. If every syllable boundary
demarcating a "stranded" consonant is available as an epenthesis
site, we obviously generate twice as many sites for epenthesis as
are necessary. This duplication, however, is needed to overcome the
weakness in Kaye and Lowenstamm's analysis that epenthesis always
occurs to the right. This is not an ad hoc duplication, but rather
there is strong empirical evidence to suggest that there are many
cases where both left and right syllable boundaries can be used as
epenthesis sites. For example, this overgeneration of epenthesis
sites is necessary to capture the acceptable variants of loan words
into English with nasal-stop clusters. Thus, for <u>Nkrumah</u> we get
both [nəkruma] and [əŋkruma]. It would be difficult to predict this
variation unless both syllable boundaries were available as epenthesis sites:

(20)   /.n. kru. ma./

Such overgeneration is also necessary if we are to predict the
acceptable pronunciation variants of a series of words in French
with optional schwa. Only if all syllable boundaries can
potentially be utilized as epenthesis sites can we get all of the
acceptable pronunciation variants.

    However, there are obviously cases in which epenthesis needs to
be restricted to one rather than the other side of a syllable
boundary. What principles can be used in restricting the operation of epenthesis processes? It is apparently true that all
languages attempt to minimize the instances of epenthesis. Thus, in
all dialects of Arabic (Broselow, 1983) in a four consonant cluster
such as :

(21)   katab + t + l + ha   'I wrote to
her'

epenthesis occurs between the first two consonants and the last two:

(22)   katabtilha

Given the restrictions against tautosyllabic consonant clusters in
Arabic, any other epenthesis site would either create an unacceptable

surface form or necessitate the introduction of an additional epenthetic epenthetic vowel:

(23)   * katab it l ha
                   * katab it ̀il ha
                   * katab t li ha

But a universal principle minimizing epenthesis is not sufficient to
predict all cases of epenthesis.

    In triconsonantal clusters, epenthesis can occur either in the
environment C____CC or in the environment CC____C. In either case,
only on application of epenthesis would it be necessary to create a
syllabic nucleus for the stranded medial consonant. For example,
Egyptian Arabic epenthesizes to the right in a triconsonantal
cluster which Iraqi Arabic epenthesizes to the left in exactly the
same environment:

(24)   katab + t + la ──▶ katabtila   Egyptian Arabic
                                                      'I wrote to
him'
              katab + t + la ──▶ katabitla   Iraqi Arabic

Broselow (1981) attempts to resolve the issue by invoking a simple
directional principle which is to be associated with all epenthesis
rules, without offering any explanation of why one dialect would
choose to epenthesize to the right while the other epenthesizes to
the left.

    Wheatley (1982:188), on the other hand, invokes a principle of
minimal allolexy to explain the difference, stating that in Egyptian
Arabic epenthetic vowels are inserted to create the least variation
in the surface form of the word:

(26)   ma + katab + t + s   'I didn't
write'

Epenthesis operates on (26) to create the surface form makatabtis,
which is an acceptable surface variant whether the following word
begins with a consonant or a vowel. Wheatley's minimal allolexy
principle is augmented by an additional principle which handles
those cases where epenthesis occurs in entirely word internal
clusters, since in these cases consideration of the phonetic shape
of the following word is irrelevant. While there is nothing in
principle wrong with invoking a variety of constraints to delimit
different instances of epenthesis, it would be theoretically
desirable if all epenthesis sites could be determined by a single
universal process.

    An additional problematic case for determining the direction of
epenthesis is Harari, a Semitic language of Ethiopia (Kenstowicz,

1978). In Harari, as in Amharic, epenthesis occurs to the right of a "stranded" consonant:

(27) $\emptyset \rightarrow i \ / \left\{ \begin{array}{l} CC\underline{\quad}C \\ \#C\underline{\quad}C \end{array} \right\}$

This epenthetic process can not be explained by Wheatley's minimal allolexy principle, since the 2 masc imperfect of 'break' is [tisäbr] while the 2 masc neg imperfect, jussive is [zatsibär].

There doesn't seem to be any single universal way to determine whether a language will epenthesize to the left or the right. However, a reexamination of Egyptian and Iraqi Arabic does suggest a possible operating principle for resolving the issue.

In Iraqi Arabic, epenthesis occurs to the left, in the first available epenthesis site. However, in order to create maximally unmarked syllables from the string in (25), a resyllabification rule will have to apply along with epenthesis:

(28)   katab + t + la $\longrightarrow$ ka. ta. bit. la

In Egyptian Arabic, on the other hand, no such resyllabification rule is necessary.

It is true that most languages epenthesize to the left, in the first available epenthesis site (Wheatley, 1982).[10] It seems probable that some languages epenthesize to the right in an attempt to minimize resyllabification in epenthesis processes. Thus, there are languages like French with a proliferation of resyllabification (enchainement) processes; however, there are languages in which resyllabification is apparently minimized. Evidence for these apparent constraints on resyllabification can be found in the types of epenthesis errors made by speakers when learning English. Broselow (1983) has pointed out that epenthesis errors by learners of English as a foreign language occur due to the speaker's effort to make syllable structure conform to the constraints in his native language. In Spanish, for example, word initial epenthesis is leftward. However, Spanish epenthesis errors in English occur rightward:

(29)   conquest   [kɔn. kə. wɛst]

             kirkwood   [kir. kə. wʊd]

If epenthesis were to occur leftward of a stranded consonant, it would necessitate a resyllabification of the word to ensure that the syllable structure conform to the constraint that onsets be maximized:

(30)   [kɔ. nək. wɛst]

              [ki. rək. wʊd]

However, the rightward epenthesis necessitates no such resyllabification. Note that this analysis also explains why inital epenthesis
is leftward even in those languages, like Egyptian Arabic and
Spanish, in which resyllabification is a highly marked process. In
initial syllables, epenthesis does not create a need for resyllabification; thus, there is no need to restrict the unmarked scansion
from left to right for epenthesis sites:

(31)   [s. la.bo]   [es. la. bo]   'Slav', Spanish

             [k. tib]   [ik. tib]   'write', Arabic

It appears that most languages epenthesize to the left in the first
available epenthesis site. Since this almost always creates a need
for resyllabification, some languages, in which resyllabification is
a highly marked process, epenthesis occurs to the right rather than
the left so that resyllabification can be avoided.

6. To recapitulate, an analysis has been proposed here which
restricts epenthesis to a single universal process operating at
syllable boundaries to create nuclear material for consonants
stranded by the operation of universal constraints on the markedness
of consonant sequences. This analysis depends solely on the
existence of syllable boundaries which are designated by markedness
principles, and explains epenthesis phenomena without resorting to
additional theoretical machinery such as the null-segment. The
ability to predict epenthesis in a universal way, and to collapse
epenthesis with the creation of syllabic consonants into a single
process are the chief descriptive advantages which ensue from the
proposals discussed here.

NOTES

[1]TS and TT represent an extension of the Cairns and Feinstein path,
but are necessary to account for languages (such as Hebrew) with
stop-fricative onsets and languages (such as Polish and Russian)
with stop-cluster onsets.

[2]Lowenstamm's second universal restriction appears to be adequate
for word internal syllable boundaries, but it is frequently violated
in word final position (e.g. English compact). There is, to my
knowledge, no satisfactory description of the constraints on word
final codas. However, nothing in the analysis presented here crucially depends on an adequate description of word final codas.

[3]There is the issue of how a language chooses betwen the two processes, but that is well beyond the scope of this paper. The
interested reader is referred to Wheatley (1982) for a discussion
of how this choice might be made.

[4]We will not be concerned here with what Broselow terms
"metrically-conditioned epenthesis." Broselow cites two examples of
this process. In Mohawk, an epenthetic vowel is inserted initially
in one syllable verbs. This is less than clear as an example of
epenthesis since the initial epenthetic vowel in Mohawk is different
from the vowel epenthesized in other positions. Broselow's other
case of metrically conditioned epenthesis is epenthesis before
initial clusters in Iraqi Arabic. Broselow claims this epenthesis
process is not governed by syllable structure constraints in Iraqi,
since it is an optional process. The status of optional epenthesis
rules will be discussed more full in Section 5.

[5]The assignment of syllable boundaries is based on Hale and White
Eagle (1980) who point out that [š] is the 2nd sg infix which can
not be associated in deep structure with the preceding syllable, and
that [šw] is not a permissable onset. (cf. šwazok ──► šawažok,
'2nd sg mash, as potatoes'.)

[6]Dinnsen (1974) views null-segments as place holders (similar to
traces in syntax) which indicate the site where material was deleted
by the application of an earlier rule. He proposes null-segments in
an attempt to constrain the operation of global rules to just those
items which contain a null-segment. However, this usage of the term
null-segment is unrelated to the usage of Anderson (1982) and Kaye
and Lowenstamm (1979) with which we will be concerned here.

[7]A similar proposal by Selkirk (1981) has stranded consonants
always serving as the coda of the "new" syllable. This was adequate
for the material from Cairene Arabic which Selkirk had under consideration. However, it is obvious that the examples Kaye and Lowenstamm use would be counterexamples to Selkirk's proposal.

[8]This vowel is variously transcribed as [ə] or [œ] The two are
phonetically indistinguishable in most dialects of French.

[9]Anderson also attempts to justify his ∅-schwa distinction by
noting that stable schwa counts in stress assignment while unstable schwa does not. However, since in a syllable based analysis the
unstable schwas can arise through epenthesis while the stable schwas
exist in the lexical representation there is no reason why the two
schwas should not have a different relationship with stress assignment.

[10]Though this principle works for the majority of cases, it does
not appear to be motivated by any theoretical requirement or depen-

dent on any regular phonological process. However, the fact that
the vast majority of iterative rules Johnson, 1968; Howard, 1972
also apply minimally and from left to right indicates that there is
probably a significant generalization in this mode of application
which our theory has as yet been unable to describe in a principled
way.

REFERENCES

Anderson, S.R. 1982 "The Analysis of French Schwa" Language
    58:534-573

Anderson, S.R. 1978 "Syllables, Segments, and the Northwest
    Caucasian Languages" in Alan Bell and Joan B. Hooper, eds.
    Syllables and Segments Amsterdam: North-Holland Publishing Co.

Broselow, E. 1983 "Non-Obvious Transfer: On Predicting Epenthe-
    sis Errors" In S. Gass and L. Selinker eds. Language Transfer
    in Language Learning, Rowley, Mass: Newbury House Publishers.

Broselow, E. (1983) "On Predicting the Interaction of Stress and
    Epenthesis" to appear in Glossa.

Cairns, C. and M. Feinstein 1982 "Markedness and the Theory of
    Syllable Structure" Linguistic Inquiry 13: 193-225.

Greenberg, J. 1966 "Some Generalizations Concerning Initial and
    Final Consonant Sequences" Linguistics 18:5-34.

Hale, K. and J. White Eagle 1980 "A Preliminary Metrical Account
    of Winnebago Accent" IJAL 46:117-132

Harris, J.W. 1969 Spanish Phonology Cambridge, . ass: MIT Press.

Hoard, J.E. 1978 "Syllabification in Northwest Indian Languages" in
    Syllables and Segments, Alan Bell and Joan Hooper, eds.
    Amsterdam:North Holland Publishing Co.

Hooper, J. 1972 "The Syllable in Phonological Theory" Language
    48:526-540.

Howard, I. 1972 A Directional Theory of Rule Application in
    Phonology Doctoral Dissertation, MIT.

Iverson, G. 1982 "The Elsewhere Condition and h-Aspire"
    unpublished paper, The University of Iowa.

Johnson, C.D. 1972 "Formal Aspects of Phonological Description" in
    Monographs on Linguistic Analysis 3 The Hague: Mouton.

Kahn, D. 1976 Syllable-Based Generalizations in English Phonology
    Doctoral Dissertation, MIT.

Kaye, J. and J. Lowenstamm 1979 "Syllable Structure and Markedness
    Theory" Proceedings of the IV G.L.O.W. Conference, Pisa, Italy.

Kenstowicz, M. 1981 "Functional Explanations in Generative
    Phonology" in Phonology in the 1980's, D. Goyvaerts, ed.   Ghent:
    E Story-Scientia.

Kenstowicz, M. and C. Kisseberth 1977 Topics in Phonological Theory
    New York; Academic Press.

Lowenstamm, J. 1981 "On the Maximal Cluster Approach to Syllable
    Structure" Linguistic Inquiry 12:575-604.

Selkirk, E. 1981 "Epenthesis and Degenerate Syllables in Cairene
    Arabic" MIT Working Papers in Linguistics.

Tranel, B. 1981 Concreteness in Generative Phonology: Evidence from
    French, Berkeley: Univ. of California Press.

Wheatley, B. 1981 Phonotactic Norms and the Prediction of
    Phonotactic Rules, Doctoral Dissertation, Indiana University.

INDEFINITE NPs AND THE INTERPRETATION OF

DISCOURSE-BASED NULL ELEMENTS*

Wynn Chao

Pennsylvania State University

In this paper I will propose that notions of markedness can be brought to bear on the question of what kinds of representations indefinite NPs (such a a man, many men, few men) should be associated within a given discourse context. These representations in turn account for the possibilities of pronominal anaphora in discourse. Crucial to this account is the behavior of what I will call discourse-based null elements. These are null elements which receive pronominal interpretation: PRO, the phonetically unrealized subject of a tenseless clause; and pro, the null subject of a tensed clause in languages that allow such constructions. Thus discourse-based null elements stand in contrast to null elements which are associated with so-called movement rules, namely WH- and NP-traces.[1]

1. NP representations and pronominal anaphora

Sentences that contain discourse-based null elements have more restrictive interpretive options than the corresponding sentences with overt pronominal elements. Consider, for example, the sentences in (1):

(1) a. Many politicians hope that they will become indispensable.

    b. Many politicians hope ____ to become indispensable.

The null pronominal in (1b) , indicated by '____', is PRO, which is interpreted as the subject of the tenseless clause 'to become indispensable'. Restricting our attention to the readings where the italicized NP and the pronominal elements are 'related' in some way, it can be seen that (1a) allows at least two readings. (1a) can be understood as meaning either that each of the relevant politicians

65

has aspirations for himself, or that many politicians hope that politicians in general will become indispensable. (1b), on the other
hand, has only one reading: that each politician has aspirations for
himself. The restriction on the interpretive options of the null
elements in general can be explained by distinguishing between the
primary, unmarked reading; and the derived, marked readings that NPs
may introduce in the context of discourse. The notion 'context of
discourse' can be made more precise via the characterization of a
discourse representation.

    I will begin by briefly laying out the structure of the discourse
representations which I am assuming, and then I will present and
argue for the distinction between marked and unmarked readings within
this system.

2. Discourse Representations

    Before going further, I should say that much of the terminology
and many of the central ideas in this and following sections derive
from the work of Hans Kamp, especially his 1981 paper ' A theory of
truth and semantic representation,' where discourse representations
are used to account for different aspects of pronominal anaphora.
However, the present account differs from Kamp's 1981 one in a
number of significant ways, discussion of which would take us away
from the main point in this paper.

    I am assuming a discourse to be defined as a sequence of dis-
course sentences: $S_1, S_2, \ldots, S_n$. A discourse sentence will be
a matrix sentence and its subcategorized sentential complements--
adverbial clauses, in-order-to clauses, if-clauses are all excluded,
and will have to be dealt with in later work. Each discourse sentence S_i is associated with a translation into some sort of logical
form S'_i. Each discourse sentence S_i is also associated with a
discourse representation DR, of which its logical form is a part.
Since each new discourse sentence changes the context of discourse
by providing additional information, it seems reasonable to allow a
new discourse representation for each sentence. Intuitively, we
want the new DR to contain the information which is still relevant
from the previous discourse representation, plus the information
which the new sentence provides. For the purposes of the present
paper, the notion of 'information' can be restricted to just the set
of discourse referents. This set defines what is available as antecedents for discourse level pronominal anaphora. I will not, then,
discuss other kinds of information (such as presuppositions, conver-
sational implicatures, propositional content, temporal linkings,
etc.) which should be included in a more developed account. (For
instantiations of such accounts, see Kamp (1981) and Heim (1982).)
Both DRs and discourse referents are being conceived of as psychological constructs; the kinds of representations that the hearer
might set up in trying to integrate the information provided by a

new sentence with the information that he already has. Thus, discourse referents should not be viewed as real, existing individuals, although very often they will correspond to real individuals.

In brief, a discourse representation DR_i induced by a sentence S_i will at least contain the following: (i) S'_i, the logical form of S_i; and (ii) a set of discourse referents U_i.

What role do the discourse referents play in the interpretation of pronominal anaphora? Given a sentence such as (2a), it is customary to say that the pronoun <u>he</u> may refer either to John or to some other individual present in the discourse context. Let us say that 'being present in the discourse context' means being a member of U, the set of discourse referents. If we additionally assume that Paul and Fred, by virtue of having been introduced in previous discourse, are the only other referents available, U for the sentence in (2a) is (2b), and the members of U are the only available antecedents for <u>he</u> in (2a).

(2) a. S: John said that he would leave.

    b. U: $\left\{p, f, j\right\}$

The set of discourse referents U_i for a discourse sentence S_i can be defined as:

(3) $U_i =_{df} U_{i-i} \cup \left\{x \mid x \text{ is a discourse referent introduced in } S_i\right\}$

In other words, U_i is the set of discourse referents already present in the discourse context, plus those introduced in S_1.[2]

Proper names such as <u>John</u> in the example above are primarily referential; and a discourse referent corresponding to them is immediately introduced into U. This introduction will take place even when the proper name does not correspond to an existing individual (for example: <u>Santa Claus</u>). But unlike proper names, indefinite NPs should not be assumed to automatically introduce discourse referents into the discourse representation. It is well known that indefinite NPs may be interpreted as either specific or as nonspecific. For example, a sentence such as (4)

(4)   Many men came in.

may be interpreted as an assertion about some specific group of men, say Tom, Dick, Harry, and Peter; or it may interpreted as an assertion about the number of men who came in. It is this second interpretation, the nonspecific one, that sould be regarded as unmarked for the indefinite. Here the NP <u>many men</u> is represented in discourse only as a quantifier in S', binding a variable in subject position

(5). No referent is introduced into the discourse representation.

(5) S'_1: MANY MEN:x [x came in]

         U_1': ∅

The specific reading is marked in that it must be derived by a discourse level rule. This rule, stated later, allows us to gather the individuals that satisfy the predicate that the quantifier has scope over, and to introduce a new referent into the discourse representation: the group consisting of those individuals. The introduction of the group referent has taken place in (6), the DR for the specific reading. Indexed lower-case g's are used to denote group constants.

(6) S'_1:   MANY MEN [x came in]

     U_1: $\left\{g_2\right\}$ (where g_2 is the discourse referent
                             corresponding to the group of men who
                             came in)

     Karttunen (1976) points out that an indefinite NP can only support pronominal anaphora in subsequent discourse if that NP is interpreted as specific. For example, consider (7):

(7) a. John didn't find a mistake.

     b. It was very easy to miss.

In (7a), <u>a mistake</u> may be interpreted as either specific or nonspecific. In the nonspecific reading, <u>a mistake</u> will be under the scope of negation and the sentence would be interpreted as saying that John found no mistakes at all. A continuation such as (7b) is not possible under this interpretation. For the pronoun in (7) to be interpreted, there must have been a specific mistake.

     In the the system presented here, this restriction on indefinites is captured in a straightforward way. A discourse sequel to (4) of the form in (8) would not be interpretable if the discourse representation associated with (4) were as in (5) above.

(4) S_1: Many men came in.

(8) S_2: They were wearing tuxedos and top hats.

In (5), there are no discourse referents that can serve as antecedents for the pronoun <u>they</u>. In (6) on the other hand, there is a possible antecedent: g_2. Thus, the continuation (8) only makes sense if the group referent has been derived and introduced into the discourse representation.[3]

3.   Primary and derived readings

One reason for arguing for the primacy of the nonspecific reading over that of the specific reading is that the specific reading
seems to be blocked in some constructions, for example, questions
and if-clauses. The (a)-sentences in (9) and (10) illustrate the
constructions in question. The italicized NPs are not understood as
referring to a particular group, hence the oddness of the (b) continuations.

(9)   a.   Did John see many students?

      b.   No, they stayed at home.

(10) a.   If most striking teachers return to work, Levesque will
           be happy.

      b.   But they are not planning to.

In (9b), they is most naturally understood as referring to students
in general, not to some particular subset of students that John was
supposed to see. In (10b), they seems to likewise refer to teachers
in general, not to some particular group: for example, those who
could be intimidated enough by newly legislated sanctions.

There are, however, no syntactic constructions where the converse
is true, i.e., no constructions which require indefinites to have
always the speciic reading. Consideration of this sort lead one to
view the nonspecific reading as primary, and the specific reading as
derived. The claim that the specific reading in indefinites is
always derived from the nonspecific one in some way is very much in
the spirit of the position argued for in Kempson and Cormack (1982).
There, they maintain that indefinite NPs must always be treated as
quantified expressions, and that 'the apparent possibility of interpretations of such indefinite noun phrases as specific has to do
with evidence justifying the truth of the proposition they are contained in rather than to do with a distinct propositional content.'
(p.37).

There is another view of indefinite NPs, which argues that they
are in fact ambiguous, sometimes having the properties of quantifiers, and other times the properties of rigid designators like
proper names. This view has been presented most recently in Fodor
and Sag (Forthcoming). It will later be argued that this second
view is incorrect. If indefinites are interpreted as either quantifiers or as rigid designators, it will not be possible to give an
adequate account of the interpretive options for pronominals in discourse.

But first, it is necessary to develop the analysis in more
detail, Following Montague (1974) and Barwise and Cooper (1981),

NPs are treated as quantifiers, composed of a determiner and a common noun. The NP $[[many]_{DET}[men]_{CN}]_{NP}$ translates as the quantifier many '(men'). The discourse rule which introduces group referents from sentences containing indefinite NPs is given as (11).

(11) <u>group referents</u>

If a discourse sentence S_i translates into a logical form S'_i of the form $[Q\hat{x}[\phi(x)]]$, where Q is a quantifier of the form $\delta(\eta)$, g_j may be introduced into DR_i; $g_j =_{df} \left\{y \mid \eta(y) \wedge \hat{x}[\phi(x)](y)\right\}$

The way in which this rule is formulated allows only those quantifiers with widest sentential scope to introduce group referents. It predicts, for example, that if (12a) is understood with <u>every profes-sor</u> taking scope over <u>several students</u> (as represented in (12b)), then <u>they</u> in (12c) cannot refer to any particular group of students talked to by the professors. (It can, however, refer to students in general.)

(12) a. S_i: Every professor talked to several students.

     b. S'_i: every'(professor')$\hat{x}$ [several'(students')$\hat{y}$ [x talked to y]]

     c. S_2:   They hadn't expected that kind of personal atten-tion.

The rule as given in (11) does not distinguish between quanti-fiers, and thus, cannot account for the fact that some quantifiers lend themselves to the specific interpretation far more readily than others: <u>many</u> CN vs. <u>most</u> CN, <u>few</u> CN vs. <u>a few</u> CN, <u>some</u> CN vs. <u>seve-ral</u> CN, etc. For example, it is easier to understand <u>many custo-mers</u> in (13a) as referring to a specific group, than it is <u>most cus-tomers</u> in (13b).

(13) a. Many customers complained about the service.

      b. Most customers complained about the service.

However, there is no absolute prohibition against a specific inter-pretation for <u>most</u> CN, since (14a) may support pronominal anaphora in (14b).

(14) a. S_i: <u>Most customers who were served by the new waiter</u> complained about the service.

      b. S_2: The manager talked to <u>them</u>, and explained the sit-uation.

The difference between (13b) and (14a) is that in (14a), the situa-tion described by the sentence combined with our knowledge of the world (that a waiter can only serve a finite number of people, and consequently, displease a finite number of people in a given time; and that a new waiter does not remain 'new' for very long) make the specific reading far more plausible. I take it, then, that the ease with which any given quantifier allows the specific reading is not something which should be captured in the formulation of the rule itself, but rather, by extrinsic factors which will determine whether the rule should apply or not in a given discourse. The intrinsic meaning of the quantifiers themselves is but one of these factors.

At which point are group referents introduced into the discourse? Since the rule is non-local, requiring for its application information about the logical form of the whole discourse sentence, it cannot apply until after that logical form has been obtained (i.e., until the translation into logical form has been completed). It follows, then, that although the representations introduced by the primary readings of NPs (discourse referents for proper names, quantified structures for indefinites) are available to pronouns within the sentence as the sentence is being translated/processed, the derived readings, and the referents thereby introduced, are not. These will only be available for pronominal anaphora after the translation has been completed. The difference in the time at which marked and unmarked referents become available in discourse will become important later.

Now consider again sentence (1a), repeated below.

(1)   a.   Many politicians hope that they will become indispensable.

One of the possible antecedents for they in (1a) is the generic class of politicians. The ability of pronouns to be interpreted as generics which have not been explicitly introduced into the discourse as such seems to be quite general, and not restricted to a few special constructions.

One way of accounting for the generic readings is by proposing another discourse rule, which will license the introduction of a discourse referent which corresponds to the denotation of the 'kind' of a common noun.

(15)   generic referents

Given a logical form S'_i containing δ , where δ is the translation of a common noun:

i.   if δ is not inside an opaque context, k_j will be introduced in the discourse representation.

ii.   if δ is inside an opaque context, k_j may optionally be introduced.

$(k_j$ corresponds to the generic class associated with the common noun translated as δ .)

In the generic reading of (1a), the NP <u>many politicians,</u> which is not inside an opaque context, induces the introduction of 'politicians' as a possible pronominal antecedent. There does not seem to be a choice: uttering or comprehending (1a) commits us to the existence of that kind of thing. When the NP is under the scope of an opaque operator, however, there is no such automatic commitment. Thus, although a generic referent can be introduced in (16) even if <u>a unicorn</u> is interpreted as nonspecific (and so, under the scope of <u>want</u>) that will not necessarily be the case, as (17) illustrates.

(16)   a. S_k: John is looking for a unicorn.

     b. S_2 Oh! Weren't <u>they</u> the fabulous beasts that could only be captured by a virgin?

(17)   a. S_1: I am looking for a nikocillium.
             (A what? A nikocillium.)

     b. S_2: You can't be serious, there is no such thing!

(17b) seems to be rejecting the existence of a kind of 'nikocillia', not only in the real world, but in the discourse representation as well.

Since the rule needs to refer to opaque contexts (which may be created arbitrarily higher up in the sentence) in order to 'know' whether clause (i) or (ii) applies, this rule is non-local as well. It too, can only apply after the translation of the sentence has been completed.

It does not seem possible to unify the rules that introduce group and generic referents. For apart from the conditions imposed by opaque contexts, the two rules differ with respect to restrictions on quantifier scope. For example, in (12), although it is not possible to introduce a group referent consisting of the students who were talked to by the professors, it is nevertheless possible to introduce the generic 'students', and understand (12b) as:

'Students don't expect that kind of personal attention.'[4]

(19) and (20) below illustrate how the system presented so far works. In order to distinguish between primary readings, which are available as the sentence is being interpreted, and derived readings, which introduce discourse referents after the logical form has been obtained, I will define, in addition to U_i, a set of U'_i:

(18) $U'_i =_{df} U_i \cup \left\{ \begin{array}{l} \text{discourse referents introduced by dis-} \\ \text{course rules applying to } S'_i \end{array} \right\}$

(19) a. S_i: John walked into his office.

U_1: $\left\{ j,\ j\text{'s office} \right\}$ his = j

U'_1: $\left\{ j,\ j\text{'s office},\ k_6(\text{'offices'}) \right\}$

b. S_2: He sat down in front of his typewriter.

U_2: $\left\{ j,\ j\text{'s office},\ j\text{'s typewriter},\ k_6(\text{'offices'}) \right\}$ he = j   his = j

U'_2: $\left\{ j,\ j\text{'s office},\ j\text{'s typewriter},\ k_6(\text{'offices'})\ k_7(\text{'typewriters'}) \right\}$

(20) a. S_1: Every professor walked into his office.

S'_1: every'(professor') $\hat{x}$ [x walked into x's office]

U_1: $\emptyset$

U_1: $\left\{ g_3(\text{the group of professors}),\ k_3(\text{professors'}),\ k_4(\text{'offices'}) \right\}$

b. S_2: ?? He sat down in front of his typewriter.

U_2: same as U'_1 he = ??   his = ??

U'_2: $\left\{ g_3,\ k_3,\ k_4,\ k_7(\text{'typewriters'}) \right\}$

I am not presently in the position to discuss the interpretation
of NPs with possessive determiners. I will simply assume that their
primary reading is referential if the pronoun is interpreted as core-
fering to a referential antecedent. In (19a) and (19b), the discourse referent j is available as an antecedent for the pronouns.
his is bound to the universal quantifier in S'_1. Bound anaphora
is strictly sentential in nature, and so operates independently of
processes which assign discourse referents as antecedents to pronouns. (However, for arguments against imposing such restrictions
on variable binding, see Lappin (1983).) (20b), although well formed
at the sentential level, is an inappropriate continuation for (20a),
since there are no discourse referents in either U_2 or U'_2 which
can serve as antecedents for he and his without a resulting feature
clash. (20b') and (20b'') are possible discourse continuations, however. In (20b'), they can be understood as either g_3 (the group
of professors) or k_3('professor'). In (20b'') they can be understood as k_4('offices').[5]

(20) b'. They are always in by 9 o'clock.

(20) (Cont.)

    b" <u>They</u> are always messy and full of books.

A consequence of the analysis presented here is that from a single syntactic occurrence of a noun phrase, multiple representation can be induced in the DR. (21)-(23) show that this must indeed be the case, since different pronouns in the same discourse can be associated with different readings induced by a single syntactic occurrence of a noun phrase.

(20)   Many guests lost <u>their</u> temper,   and <u>they</u> had a big fight.
                     bound variable   group

(22)   Few congressmen made   <u>their</u>   views on the issue public, and
                         bound variable
    <u>they</u> were surprised at the lack of reaction from the press.
    group

(23)   John owns many parrots. <u>They</u> are tropical birds, and cannot
                               generic
    withstand the Canadian winter. He must keep <u>them</u> in a climate-
                                              group
    controlled greenhouse.

4. Compositional meaning, context dependent interpretation

The two readings of sentence (1a) can be represented by associating the pronoun <u>they</u> with different representations of the NP <u>many politicians</u>.

(24)   S: Many politicians hope that they will become indispensable.

Bound variable reading

(25)   S': many'(politicians') $\hat{x}$[xhopes that x will become indispensable]

    U: $\emptyset$

    U': $\left\{ g_2(\text{the group of politicians}), k_2(\text{'politicians'}) \right\}$

Generic reading

(26)   S':   many'(politicians') $\hat{x}$[x hopes that they will become indispensable]

    U: $\emptyset$

(26) (Cont.)

U': $\left\{ g_2(\text{the group of politicians}), k_2('\text{politicians}') \right\}$

they = k_2

Most speakers seem to disallow the group reading for this sentence. In the present account, it means that they = g_2 must be ruled out for the cases in which the pronoun is in the scope of the quantifier to which the group rule applies. g^2 for (25) is (27); for (26) is (28).

(27) $g_2 = \left\{ y \mid \text{politicians}'(y) \wedge \hat{x}[\text{ x hopes that x will become indispensable}] \text{ } (y) \right\}$

(28) $g_2 = \left\{ y \mid \text{politicians}' \text{ } (y) \wedge \hat{x}[\text{ x hopes that they will become indispensable}] \text{ } (y) \right\}$

In (27), they cannot be g_2 because it has already been interpreted as a bound variable before g_2 could be defined. In (28), a similar situation obtains. They need also have been interpreted, otherwise it will be impossible to establish which individuals belong to the group.

In contrast to (1a), (1b), repeated below, only allows the null element to be interpreted as a bound variable, as in (29b).

(29)   a.   S: Many politicians hope _____ to become indispensable.

  b.   S':   many'(politicians') $\hat{x}$ [x hopes that x will become indispensable]

  c.   U: $\emptyset$

  d.   U': $\left\{ g_2, k_2 \right\}$

Nothing in the discourse rules can prevent the introduction of the discourse referents in U' (29c), but their association with PRO is blocked. One could account for this restriction by stipulating that PRO can only be associated with the unmarked reading of the NP, in this case, the quantified reading. But this stipulation does not explain, for one thing, how it is possible to distinguish the marked and unmarked readings in U', and moreover, the explanation will not generalize adequately to a wider range of data.

I would like to explore another solution here. This solution depends crucially on the fact that marked readings become available in discourse later than the unmarked ones. To see why this is relevant, consider first that it is possible to view the interpretation

of sentences in discourse as being done in two stages. The first
stage is the procedure by which the <u>compositional meaning</u> of the
sentence is derived. This is obtained by combining the inherent
meaning of the basic constituents of the sentence (for accounts on
how this procedure may actually be carried out, see Montague (1974),
Partee (1975b), Dowty, Wall and Peters (1981), to name a few sources). Alternatively, we could envisage this first stage as the
procedure which is a representation similar to a 'logical form' in
the Government Binding framework. In any case, I am assuming that
some intermediate level of representation is the output of this first
stage,--a 'logical form', or a formula in intensional logic--and
that this output is not fully interpreted until the second stage.

The second stage, <u>context dependent interpretation</u>, supplies the
reference to pronominals which have not yet been interpreted, and in
general supplies the kinds of information that could not have been
obtained independently of the context within which the sentence is
produced.

Overt pronouns have inherent meaning. Not only do they contribute information about grammatical category and grammatical function
to the sentence, they also contain information about gender, number,
and person. So they enter into the processes which derive the compositional meaning of the sentence in the same way that any other
lexical items do. From a processing point of view, however, there
is no reason to assume that null elements have any inherent features
in the same sense that overt pronouns do.[6] Thus, null elements are
not able to contribute to the compositional aspect of the interpretation of the sentence unless they themselves can first be interpreted somehow. This can be done sentence-internally, by interpreting
the null element as a variable bound by a quantified NP (as in (25)
and (29)), or it can be done contextually, by interpreting the null
element as coreferential to some available discourse antecedent. If
either interpretation has taken place, the inherent features of the
antecedent are 'taken over' by the null element, allowing it to contribute to the construction of the compositional meaning of the sentence. But the fact that the null element must be interpreted before
the first-stage interpretation may be completed means that the only
discourse referents available to a null element will be either those
referents already introduced in previous discourse representations,
or those introduced by the primary, unmarked readings of NPs in the
same sentence. In other words, only those discourse referents that
are present in U. The marked discourse referents (those introduced
at U') may not be introduced until <u>after</u> first-stage interpretation
has been completed, and <u>that</u> cannot be completed, until after the
null elements have been interpreted.

Applying what has been said to example (29), it can now be said
that the bound variable reading is the only reading available for
PRO, the null element, because the quantifier is the only representation introduced by the NP that is available at the point that PRO

must be interpreted. The particular example under discussion is, perhaps, not the best illustration for this point. (29) has been argued to be a structure of <u>obligatory control</u>, which required that PRO be bound to the subject of the higher clause. This would automatically preclude PRO being interpreted as coreferential to an antecedent introduced by earlier discourse. Examples presented in (30), however, will show that the account presented here does describe the facts correctly. But first, turning back to the constructions where the overt pronoun is involved, notice that overt pronouns need not be contextually interpreted before compositional meaning is computed. The result is that they can wait longer than null elements for contextual interpretation. As a consequence, the discourse referents in U' are also available as antecedent for overt pronouns (as in (26)), assuming, of course, that the interpretation does not result in as inconsistent assignment of features such as gender, number, etc.

The sentences in (30) contain structures of nonobligatory control (in the sense of Williams (1980))--i.e., the "controller" for the null element is not uniquely determined by the syntactic construction or the higher predicate. But even then, the restriction on the interpretation of PRO stills holds.

(30) a. Many politicians told Bill that it would be hard _____ to resign.

      b. The need _____ to become indispensable is found in many politicians.

In (30a), the controller for the null element may be understood as either <u>Bill</u>, or the quantifier <u>many politicians</u>, which binds the null element. It cannot be understood as 'the group of politicians', or the generic 'politicians'. Another possible reading for (30a is the <u>arbitrary reading</u>, 'hard for anyone to resign'. This reading also does not depend on discourse referents introduced at U'--we might assume, for example, that the set of discourse referents is never really empty, that there will always be an 'arbitrary' referent available, with various conditions on when it can be used as an antecedent. Similarly, (30b) may be understood as expressing an arbitrary reading (The need for one to become...), or a bound variable reading (Many politicians need _____ to become...). The derived readings are absent. But when overt pronouns are used in the same constructions, these readings become available again:

(31) a. Many politicians told the reporter that it would be hard for <u>them</u> to regain the confidence of the public after the scandal.

      b. The need for <u>them</u> to regain public confidence after the scandal drove many politicians to seek press conferences to clear their names.

5. Null subjects

Even more striking is the fact that in at least one Pro Drop language, Brazilian Portuguese, the null subject in tensed clauses obeys the same restriction on the interpretation of the null element as the null subject in tenseless clauses. The null subject in tensed clauses is often claimed to behave in the same way as an overt pronoun, and thus there is no a priori reason to expect that it should share this restriction with PRO. So it is interesting to see that this is in fact the case, as (32) shows.[7]

(32)   a. Muitos políticos pensam que <u>eles</u> são indispensáveis.
                                        bound variable
                                        generic

       b. Muitos políticos pensam que _____ são indispensáveis.
                                        bound variable
                                        * generic

       'Many politicians think that (they) are indispensable.'

       c. Muitos políticos pensam _____ serem indispensáveis.
                                   bound variable
                                   * generic

       'Many politicians think PRO to-be$_{pl}$ indispensable $_{pl}$.[8]'

(32a) and (32b) contain tensed complements with and without overt subjects, respectively. (32c) contains the tenseless complement with the inflected infinitival. <u>Pensar</u>, 'to think' in Portuguese, is similar to <u>hope</u> and <u>believe</u> in English in allowing both tensed and tenseless complements. It is unlike them, however, in that the null subject in the inflected infinitival clause need not be controlled by the higher subject. (33), where the higher subject does not control, is grammatical:

(33)   João pensa _____ serem indispensáveis.

       'João thinks [they] to-be$_{pl}$ indispensable$_{pl}$.'

Another thing that should be pointed out is that the judgments in (32a) - (32c) cannot be captured by stipulating that null elements cannot be associated with derived readings. Derived readings associated with NPs introduced in earlier discourse sentences are available to null subjects, as (34) shows.

(34)   S$_1$:   Muitos turistas se queixaram da comida.

       S$_2$:   Mas isso não surpreende, porque geralmente _____ ficam

               insatisfeitos com a comida estrangeira.

'Many tourists complained about the food. But this is
not surprising, since [they] are usually dissatisfied
with foreign food.'

At U'$_1$, discourse referents corresponding to the class of tourists
(generic reading), and to the group of tourists who complained have
been introduced. These referents are then available to the null pronoun in S$_2$, as is expected given the two stage interpretation of
sentences in discourse being advocated here.

The data involving null subjects also argues that indefinite NPs
always have a primary, quantificational representation, and argues
against treating indefinites as ambiguous, either quantifiers or referential expressions. If indefinites were ambiguous, no reading
should be more basic than another, and there can be no explanation
fo the unavailability of the group (=referential) readings for the
sentences in (32). Null subjects also support the claim that NPs
can introduce multiple representations in discourse, as (35) shows:

(35)   S$_1$: Muitos políticos pensam que _____ são indispensáveis.
                                              bound variable
          S$_2$: Mas _____ estão muito enganados.
                  group reading
                  generic reading.

          'Many politicians think that [they] are indispensable.
          But [they] are very mistaken.'

In (35), the indefinite NP must function as a quantifier in order to
bind the variable in S$_1$, but it must also function as a referring
expression, in order to provide an antecedent for the null element
in S$_2$. Under the ambiguity hypothesis, the indefinite NP would
have to be represented as either one or the other, leaving one of
the pronominal interpretations unaccounted for.

6. Deictic readings

The analysis presented here can also explain an otherwise puzzling fact about null subjects. It is well known that they can refer
to referents which are prominent in the discourse context, even when
these referents have not been linguistically introduced. In this
respect, they behave like pronouns.

(36)   Scenario: John walks into a room, and all eyes turn
          towards him. Someone may say either (a) or (b).

          a.   _____ está muito bem vestido hoje.

          b.   Ele está muito bem vestido hoje.

          '(He) is very well dressed today.'

It seems reasonable to assume that among the rules that introduce
discourse referents, there is one which introduces referents corresponding to contextually salient entities. If so, j will be present in U of (36), and available as an antecedent for the null subject. The puzzling fact is, however, that although null subjects
can refer to contextually introduced referents, they can never be
used deictically. If instead of the scenario in (36) we had (37):

(37) Scenario: Three men walk into the room at the same time. All
      three are equally salient, but only one of them, John, is well
      dressed.

      a. ?? _____ está muito bem vestido hoje.

      b. Ele está muito bem vestido hoje.

Under those circumstances, it would be odd to omit the pronoun, even
though it is quite clear that only John could be the intended referent. (37b) is fine, with ele understood as deictic. One could think
of deictic acts as instructions to introduce referents into a discourse representation. This, then, is what the deictic ele does.
As the three men walk in in (37), a group referent is introduced
into the DR by the same process that introduced j for (36a) and
(36b); but group referents do not automatically make the members in
that group available as individual antecedents. The deictic pronoun
in (37b) introduces j into the associated DR. But a null element,
which has no inherent features or properties, cannot effect such an
introduction--it needs to be interpreted first. But its interpretation will also make it either bound or coreferential to something
else in the discourse. As a result, a null pronominal element cannot
possibly be deictic.

7. Conclusion

    Many aspects of this proposal still need to be developed. For
example, disjoint reference has not been dealt with at all, and
nothing has been said about the fact that some 'available' discourse
referents are more available than others (see, for example, Dahl and
Gundel (1981) for discussion on the importance of the notion 'topic'
in the determination of antecedenthood of nonfocussed and null pronominals; see also Bach (1982) for discussion on the kinds of more general considerations involved in the determination of the controller
in cases of 'free' control). It would also be desirable to establish
that the null elements that arise as a result of movement rules
('gaps', or 'traces') behave in similar fashion. Minimally, it would
be nice to show that traces need to be interpreted before the initial
processing of the sentence can be completed. Mailing and Zaenen
(1982) suggest that this may be the case--that gaps lead to an in-
terruption in the initial processing of the clause while resumptive
pronouns don't. Resumptive pronouns are linked to their antecedents

via a later, different process from that which interprets gaps. An explicit statement about how this proposal should be integrated with existing frameworks is also desirable, but goes beyond the scope of the present contribution.

What I have done in this paper is argue for the following points:

(i)   A single occurrence of a NP may induce multiple representations into the Discourse Representaion. These representations can be distinguished into primary and derived, the primary one being irreducible and immediately available, while the derived ones are available only after translation into some logical form has been completed.

(ii)   Null discourse-based pronominal elements have no inherent meaning; any features that can be attributed to them are derived from contextual interpretation. I have assumed, but not explicitly given, a system in which there are two aspects to the interpretation of sentences in discourse: compositional and discourse dependent. Because of their lack inherent meaning, null elements must be interpreted before the completed logical form for the sentence can be obtained. This restricts their choice of antecedents to either referents introduced in earlier discourse, or unmarked representations introduced by NPs in the same discourse sentence. This proposal predicts that the same restrictions hold for PRO and the null subject in tensed clauses, even though the null subject has been argued to behave much more like an overt pronoun that PRO does.

Footnotes

* I would like to thank Emmon Bach, Denis Bouchard, Marcia Carlson, Robin Cooper, Gennaro Chierchia, Helen Goodluck, Hans Kamp, Ruth Kempson, Shalom Lappin, Esmeralda Vailati Negrão, João Andrade Peres and Mats Rooth for helpful discussion and comments. I must also thank Mark Feinstein, who in discussing my earlier work would not allow generic readings of pronouns to be dismissed as irrelevant phenomena. Any errors and omissions remain my sole responsibility.

[1] It is not in any way crucial for the present analysis that pronominal null elements be syntactically represented as an NP dominating no lexical material. What is crucial is that they be represented in the logical form, in whatever way the notion 'subject of a clause' needs to be formally represented.

[2] This formulation is adequate for the purpose of this paper. However, it will have to be modified if the present analysis is extended to opaque contexts, since NPs within these contexts do not necessarily give rise to discourse referents.

[3]The group reading has been first discussed, to my knowledge, in Evans (1980). See also Higginbotham (forthcoming).

[4]Robin Cooper (personal communication) suggests, however, that the restriction to 'highest quantifier' in rule (11) may follow from independent considerations in the system presented in Barwise and Cooper (1981).

[5]It may also be possible to get an interpretation of _they_ as 'professor's offices.' This suggests that the rule which introduces generic referents should be modified to allow also more restricted 'kinds' of generics.

[6]A distinction should be drawn between a formal characterization of null elements which allows some to contain inherent features (cf. proposals concerning PRO in the GB framework (Chomsky 1982)), and the question of whether formally assigned inherent features are actually available to the construction of the logical form as the sentence is being initially interpreted.

[7]The fact that null subjects in both tensed and (inflected) tenseless clauses cannot access antecedents at U' indicates that at least for Portuguese, overt inflectional marking on the verb is not sufficient to allow inherent features of the subject to be recovered from verbal inflection alone. Thus, it argues against the position often expressed in the GB framework, that null subjects are in some sense licensed by AGR, because the latter's features allow the null subject to be interpreted.

[8]Sentences such as (i)-(lv) have been pointed out, which might be taken to show that a group reading is available to pronominals in the the scope of quantifiers which serve as their antecedents.

(i) Many youngsters decided that they could overturn a truck.

(ii) Many youngsters decided _____ to overturn a truck.

(iii) Some students asked if they could do their homework together.

(iv) Some students asked _____ to do their homework together.

At issue here is whether the pronominal (overt or null) in these examples refers to the group defined by the discourse rule (11). It seems clear that the pronominal must be interpreted as a plural or collective entity. Consider (i): this sentence most likely does not mean that each of the youngsters decided that he was capable of overturning a truck on his own: it means that there was some kind of collective decision. Likewise, (iv) only makes sense if the students intend to do their work in a group.

I would like to suggest that the plural readings in these examples are not group readings derived by discourse rule (11), but rather, the result of the quantifier binding to a plural variable. It is normally assumed that individual variables range only over single individuals. But it does not seem unreasonable to suppose that plural pronouns which are bound actually may be able to take groups of individuals as values. One test for distinguishing between a discourse level group reading and binding to a plural variable has been suggested by Mats Rooth (personal communication). Since rule (11) introduces a single group referent into the DR, uniqueness of reference for the pronoun is the result. In (v),

(v)   Many youngsters were out in the street last night. <u>They</u> overturned a truck.

a single group of youngsters overturned (most likely) a single truck. A bound variable, on the other hand, may take on multiple values, since it is dependent on the quantifier for its interpretation. (vi) and (vii) show that to be true for pronominals in these constructions:

(vi)   During the riots, many demonstrators decided that they would overturn a government truck.

(vii) Many students asked if they could leave in groups.

It does not seem difficult to interpret (vi) as meaning that several trucks were overturned, each by a number of different demonstrators. And (vii) only makes sense if the intention is that the students divide up into a number of smaller groups. As for the cases in (i)-(iv), it may be that other factors are at work in restricting the interpretation of the sentence in such a way that the quantifier is only satisfied by a single plural value.

References

Bach, Emmon. 1982. Purpose clauses and control. <u>The nature of syntactic representation</u>, ed. by Pauline Jacobson and Geoffrey K. Pullum, 35-57. Dordrecht: Reidel.

Barwise, Jon and Robin Cooper. 1981. Generalized quantifiers and natural language. <u>Linguistics and Philosophy</u> 4.159-219.

Chomsky, Noam. 1982. <u>Lectures on government and binding</u>. 2nd ed. Dordrecht: Foris.

Dahl, Deborah and Jeanette Gundel. 1981. Identifying referents for two kinds of pronouns. Paper presented at the LAS Annual Meeting, New York.

Dowty, David R., Robert E. Wall and Stanley Peters. 1981. <u>Introduction to Montague Semantics</u>. Dordrecht: Reidel.

Evans, Gareth. 1980. Pronouns. <u>Linguistic Inquiry</u> 11.337-362.

Fodor, Janet D. and Ivan Sag. Forthcoming. Referential and
    quantificational indefinites. Linguistics and Philosophy.
Heim, Irene. 1982. The semantics of definite and indefinite NPs.
    Ph.D thesis, University of Massachusetts at Amherst.
Higginbotham, James. Forthcoming. Some remarks on binding theory
    and logical form. Linguistic Inquiry.
Kamp, Hans. 1981. A theory of truth and semantic representation.
    Formal methods in the study of language, ed. by J. Groendijk, T.
    Janssen and M. Stokhof, 277-322. Amsterdam: Mathematical Centre
    Tracts (#'s 135, 136).
Karttunen, Lauri. 1976. Discourse Referents. Syntax and Semantics 7,
    363-385. New York: Academic Press.
Kempson, Ruth and Annabel Cormack. 1982. On specificity. ms.,
    University of London.
Lappin, Shalom. 1983. VP anaphora, quantifier scope and Logical
    Form. ms., University of Ottawa
Maling, Joan and Annie Zaenen. 1982. A Phrase Structure account of
    Scandinavian extraction phenomena. The nature of syntactic
    representation, ed. by Pauline Jacobson and Geoffrey K. Pullum,
    229-282. Dordrecht: Reidel.
Montague, Richard. 1974. The proper treatment of quantification in
    ordinary English. Formal philosophy, ed. by Richmond H. Thomason,
    245-270. New Haven: Yale University Press.
Partee, Barbara H. 1975a. Deletion and variable binding. Formal
    semantics of natural language, ed. by Edward Keenan. Cambridge:
    Cambridge University Press.
Partee, Barbara H. 1975b. Montague Grammar and Transformational
    Grammar. Linguistic Inquiry 6.203-300.
Williams, Edwin. 1980. Predication. Linguistic Inquiry 11.203-238.

MARKEDNESS, GRAMMAR, PEOPLE, AND THE WORLD*

Bernard Comrie

University of Southern California

0. Introduction

In the study of markedness, there are two possible approaches to explaining markedness. On the one hand, one could simply say that markedness is a formal property of grammars, at best to be explained as part of the panhuman species-specific genetic inheritance of language users as members of the species homo sapiens. On the other hand, one could try to account for markedness in terms of other, independently verifiable properties of people, the world, or people's conception of the world. In this paper, I shall be arguing that, at least in a large number of instances, this second approach provides a viable explanation for observed markedness facts, including crucially a number of instances where there is no formal reason to assume that the markedness facts would fall out the way they do: either the opposite distribution of markedness would be equally simple formally, or some other distribution of markedness would be considerably more simple formally.

Although most of my examples will be from syntax and morphology, I will illustrate this initially with an example from phonology. From the marking conventions given by Chomsky and Halle (1968:403-414), one can deduce the generalization that sonorant segments are voiced in the unmarked case, while nonsonorant segments are unvoiced in the unmarked case, i.e.

(1) [u̲ voice]   [α voice]   /   [⁻⁻son⁻⁻]

From a purely formal viewpoint, there is no reason to expect   in (1) rather than - . Once, however, we look at the phonetic characterization of 'sonorant', the correlation between sonorant and voice becomes clear:

85

(2)   'Sonorants are sounds produced with a vocal tract
       configuration in which spontaneous voicing is pos-
       sible; obstruents are produced with a cavity con-
       figuration that makes spontaneous voicing impossible'
       (Chomsky and Halle, 1968:302).

The examples treated in the body of this paper come from various
domains of syntax and morphology, but all share one property in
common: In each case, I argue that the construction type which is
least marked formally is also least marked in terms of properties of
the real world, or more accurately in terms of people's conception
of the world; hence the inclusion of 'people' and 'the world' in the
title of this paper.

By 'formally unmarked' I mean throughout the absence of formal
marking, while 'formally marked' refers to the presence of such marking; 'degrees of markedness' thus refer to different degrees of
explicitness of overt marking. Although there are surely cases where
lack of formal marking does not match unmarkedness within the system,
there is a strong correlation between the two (Greenberg 1966:26-27),
and I believe that this correlation holds in the examples discussed
below.

In section 1, I treat an example from Armenian morphology which
provides a particularly clear illustration of the methodology to be
used throughout the paper. The examples in sections 2 through 4 are
more complex, but I believe also more telling, especially in that
they bring together data from various languages.

1.   Armenian locatives

In Modern Eastern Armenian, there are in principle three ways of
expressing location: the citation form of the noun phrase, as in
(3); the locative case in -um, as in (4); and use of a postposition,
most of which require the genitive case of the noun, as in (5):

(3)   Aprum   em   Yerevan.

       living I-am Erevan

(4)   Aprum   em   Yerevan-um.

       living I-am Erevan-LOC

(5)   Aprum   em   Yerevan-i meǰ.

       living I-am Erevan-GEN in

       'I live in Erevan.'

In terms of degrees of formal markedness, the bare citation form is

the least marked (it carries no overt marking of location); the locative case occupies an intermediate status, since it does overtly
indicate location, but does not specify further the kind of location
involved; the postpositional construction is the most marked, since
it specifies the precise kind of locational relation involved (contrast mej 'in with vəra 'on').

Although the three constructions are often interchangeable, as in
(3)-(5), with minimal differences among the alternatives,[2] this is
not always the case. The choice among the three possibilities involves a correlation between the formal markedness of the locative
construction and the degree of markedness of the locational situation in the world being described. The unmarked situation is where
one has a simple statement of location, i.e. a certain entity is
said to be located at a certain place, as in (3)-(5). Only in this
constellation of locational verb and noun phrase of place is construction type (3) fully acceptable.[3] Thus, if one replaces the
verb aprel 'to live' with utel 'to eat', one gets the marginal sentence:

(6)   ?Utum   em   Yerevan.

      eating I-am Erevan

      'I eat in Erevan.'

Preference here is given to Yerevan-um or Yerevan-i mej

Turning to the difference between the locative and the postpositional construction, and recalling that the locative expresses only
the most general notion of location, we find that the locative is
preferred with noun phrases referring to places, and is interpreted
as the most natural (least marked) locational orientation between
entity located and place of location. For a city, this is 'in', as
in (3); for a street, it is 'on', since Armenian, like American
English, conceptualizes location as being on, rather than in, a
street:

(7)   Aprum   em   ays   pʰoɣocʰ -um/pʰoɣocʰ-i   vəra.

      living I-am this street -LOC street-GEN on

      'I live on this street.'

For noun phrases that are not specifically names of places, but
which refer to entities that can readily be conceived of as places
(i.e. typically noun phrases referring to inanimate objects locatable
in three-dimensional space), the locative is still usable, with the
interpretation of the most natural locational orientation, although
here the postpositional construction is often preferred. Suppose,

for instance, that we want to locate a pin relative to a box. This can be done using the locative, but only if the pin is in the box: since a box is a receptacle, the most natural locational orientation with respect to it is location inside of, rather than, say, on top of. The latter could only be expressed using the posposition vǝra 'on'.

(8)   Gǝndaseɣ-ǝ tup^h-um e.

      pin-DEF   box-LOC   is

(9)   Gǝndaseɣ-ǝ tup^h i meǰ^{v} e.

      'The pin is in the box.'

(10) Gǝndaseɣ-ǝ tup^h -i vǝra e.

      pin-DEF   box-GEN on   is.

      'The pin is on the box."

In idiomatic usages, where no contrast in locational orientation is possible, the locative is preferred:

(11) Hodvac-n   ays gǝrk^h-um e.

      article-DEF this book-LOC is

      'The article is in this book.'

In (11), there is no possible contrast between an article being in as opposed to on a book, so the locative is preferred (although gǝrk^h-i meǰ$_v$ is not excluded). This may be contrasted with someone keeping a flower in a book, an unnatural content-receptacle relation, where the postpositional construction is preferred:

(12) Caɣik-n   ays gǝrk^h-i meǰ^{v} e.

      flower-DEF this book-GEN in is

      'The flower is in this book.'

Finally, with animate noun phrases as place of location, only the postpositional construction is possible, as in the following example from Minassian (1980:333):

(13) Ays   avazak-i   meǰ^{v} mi k^{h}ani lav   hatkut^{h}yunner kan.

      this brigand-GEN in some   good qualities   there-are

      'There are some good qualities in this brigand.'

In (13), the locative avazak-um would be simply ungrammatical. This correlates with the fact that animate beings are the most difficult to envisage as places, and also emphasizes, incidentally, that the relevant parameter is people's conceptualization of the real world, rather than actual properties of the real world: physically, animate beings make just as good receptacles, or locational orienters, as inanimate objects, but it turns out that people do not think of animate beings in this way.

To summarize the discussion of section 1: For the least marked locational situation, where an entity is simply located in a certain place, Armenian can use the formally least marked construction (bare citation form of the noun phrase). For situations intermediate in markedness, i.e. where an entity or event is located in terms of the most natural orientation with respect to an object, especially where only one orientation is possible or the object is readily conceivable as a place, the formally intermediately marked construction is used, namely the locative in -um. For the most marked situation, where location is either in terms of a less natural orientation with respect to an object, or is with respect to an entity not easily conceivable as place, only the most marked construction is possible, with a postposition.

2. Control

Under control here, I understand the principle or principles which determine the interpretation of the missing subject of the infinitive construction in examples like (14) and (15):

(14)   Tom intends to return before nightfall.

(15)   Tom persuaded Sally to return before nightfall.

In (14), the understood subject of to return before nightfall is the matrix clause subject, Tom; in (15), the understood subject of the infinitive is the matrix clause object, Sally. From the early generative literature, two lines of explanation have been suggested for the difference between subject-control in (14) and object-control as in (15). One is in terms of the formal structure of the sentences, in particular the closeness in syntactic structure between the trigger (i.e. the matrix clause noun phrase appearing overtly in the sentence) and the target (i.e. the noun phrase that is understood, but not expressed overtly in the sentence); this 'minimal distance principle' is proposed, under the name 'erasure principle', by Rosenbaum (1967). The second is in terms of the semantics of the two constructions, and was introduced into the generative literature by Postal (1970:468-476); it runs along the following lines: intentions are normally (i.e. in the unmarked case) with respect to one's own actions, whereas persuasion is normally with respect to actions of the person being persuaded; thus the control properties of the verbs in (14) and (15) simply use the formally least marked construction,

with omission of the subject noun phrase, for the situation that is
conceived as most natural. For more marked situations, i.e. an intention relating to someone else's actions, or persuasion not relating to an action of the person being persuaded, it is necessary
overtly to mention the subject of the embedded clause:

(16)   Tom intends that Sally should return before nightfall.

(17)   Tom persuaded Sally that he (i.e. Tom) should return
        before nightfall.

Example (17) illustrates that mere coreferentiality between some
matrix noun phrase and the embedded subject is insufficient for
omission of the embedded subject.) I believe that a full account
of control will require both formal principles and a concept of
markedness related to markedness of (people's conception of) situations. For the purposes of the present paper, I will concentrate
on the relevance of markedness related to nonlinguistic properties
of situations, although reasons for considering that this interacts
with, rather than supersedes, formal principles will be indicated in
passing.

One problem for the minimal distance principle is the set of
control properties of the verb promise and its close synonyms:

(18)   Tom promised Sally to return before nightfall.

The interpretation of (18), at least for most speakers of English,
is that it refers to Tom's returning before nightfall, not Sally's,
in violation of the minimal distance principle.[4] In terms of the
alternative approach to control, however, sentences with promise turn
out not to be exceptional. In order to justify this conclusion, it
will be necessary to make more explicit precisely what is implied by
relating the interpretation of sentences like (17) and (18) to properties of people's conceptualization of the world.

Searle (1969:54-71) sets up necessary-and-sufficient conditions
for the definition of a number of speech acts, of which the ones relevant to our current discussion are directives (referred to there
as requests) and promises. The class of directives subsumes the set
of verbs that occur in constructions like (16), with object control.
The crucial condition for our purposes is the preparatory condition
for directives, namely that the speaker should believe that the
hearer is able to carry out the act referred to by the directive (in
our examples, by the embedded clause). Linguistically the least
marked construction to refer to an entity being able to carry out an
act is to express that entity as subject of the corresponding linguistic construction; moreover, the least marked construction linguistically for referring to the hearer (recipient) of a directive is to
express that hearer as object of the matrix verb which indicates that

a directive is involved. Thus, in order to express a situation where Sally is recipient of a directive from Tom and is moreover the person who is to carry out that directive – the least marked directive situation – we would expect to find <u>Sally</u> as both matrix object and embedded subject. The possibilities for omitting noun phrases after <u>persuade</u> in English (and many other languages) allow one to omit the noun phrase, i.e. use the less marked construction formally, precisely where one has this coreference of matrix object and embedded subject, which correlates with the least marked situation.

For promises, on the other hand, the crucial condition is the propositional condition, whereby the promise must refer to a future act of the speaker's. The most natural linguistic reflection of this is a clause referring to the promise in which the speaker appears as subject of that clause. Thus, from the pragmatics of promises, we would expect to have most naturally coreference between promiser and the person to carry out the promise, i.e. between matrix subject and embedded subject; and this is precisely what we find in sentences like (18). Thus, the distinction between (17) and (18), inexplicable on purely formal grounds, does have a ready explanation in terms of the pragmatic differences between directives and promises.

At first sight, sentences like (19)-(24) might seem to pose a problem for the pragmatic explanation, since the entity required to carry out the act is not the recipient of the directive or the originator of the promise:

(19)   Tom persuaded Sally that her (i.e. Sally's) son should attend public school.[5]

(20)   Tom promised Sally that his (i.e. Tom's) son would attend public school.

(21)   Tom persuaded Sally that he (i.e. Tom) should return before nightfall.

(22)   Tom promised Sally that she (i.e. Sally) would return before nightfall.

(23)   Tom persuaded Sally to be interviewed by Walter Cronkite.

(24)   Tom promised Sally to be interviewed by Walter Cronkite.

Sentences (23)-(24) would seem to be particularly damaging, since in these passive infinitive constructions the recipient of the directive and the originator of the promise apparently undergo the interviewing process, rather than carrying out any action: this would suggest a formal principle (only derived subjects can be omitted) rather than a pragmatic principle (although this formal principle would still leave unexplained the difference between the control

properties of the matrix verbs in (23) and (24)). However, it should
be noted that there is one crucial aspect of interpretation that is
common to all of (19)-(24): in each example, the recipient of the
directive or the originator of the promise must be interpreted as
having some degree of influence over the event referred to by the
embedded clause. Thus, (19) only makes sense if Sally has influence
over whether or not her son attends public school; (20) only makes
sense if Tom has influence over what his son does; (21) implies
that Sally has influence over whether Tom returns; (22) that Tom
has influence over whether Sally returns; (23) that Sally has influence over whether she submits herself to an interview by Walter
Cronkite; and (24) that Tom has such influence.[6]

Further evidence in favor of the pragmatic approach can be found
in other idiosyncrasies of the promise class of verbs. In general,
when the matrix verb of an infinitive construction is passivized,
the truth-functional meaning of the sentence is unchanged, as in the
following pair:

(25)   Tom persuaded Sally to return before nightfall.

(26)   Sally was persuaded by Tom to return before nightfall.

In both (25) and 26), Sally is understood as subject of the infinitive. With promise, passivization of the matrix verb generally
leads to results of highly marginal grammaticality:

(27)   ?*Sally was promised by Tom to return before nightfall.

For reasons that I do not fully understand, when speakers who interpret (18) above to mean only 'that Tom would return before nightfall'
are forced to assign an interpretation to (27), the interpretation
usually assigned is 'that Sally would return before nightfall'.[7]
If, however, this interpretation is made formally overt by including
in the infinitive construction the expression to be allowed to, then
the resulting sentence is, for me and many other speakers of English
who share the judgments given above on (18) and (27), fully
acceptable:

(28)   Sally was promised by Tom to be allowed to return before
        nightfall.

In (28), the omitted derived subject of to be allowed to return is
coreferential with Sally. Formally, there is no reason to expect
this difference between (27) and (28) simply because of the insertion of to be allowed to, but pragmatically the leap in acceptabi-
lity from (27) to (28) is readily explainable: (28) makes explicit
that the omitted subject is the recipient of permission to do something, and in terms of promises it makes much more sense (is more
natural, is the the unmarked situation) that the receiver of a

promise should thereby receive permission to do something, rather
than that the giver of a promise should undertake to receive permission to do something (which latter approaches pragmatic incoherence). Indeed it is possible that (28) is interpreted directly in
terms of its pragmatics, rather than being mediated through the
partial interpretations provided by its formal constituents. Finally, it may be noted that the matrix active correspondent to (28),
for me, is much more readily given the interpretation where Sally is
understood subject of the infinitive than would be possible in (18),
although (29) is considerably less acceptable in this interpretation than is (28):

> (29)   ?Tom promised Sally to be allowed to return before
nightfall.

(The question mark relates to the interpretation where Sally is
allowed to return.)

Further evidence in this domain is provided by the matrix verb
ask, which in English covers both the interpretation 'request' and
the interpretation 'request permission for'. Sentence (30) below
has, for all speakers, the dominant interpretation 'request', i.e.
'John asked of Mary that Mary leave the room':

> (30)   John asked Mary to leave the room.

In this sense, ask behaves as a regular transitive matrix verb (by
the minimal distance principle) or as a regular directive verb (by
the pragmatic principle) For some speakers, there is a secondary,
less preferred, interpretation of (30) as 'John asked Mary for
permission that he (John) might leave the room', where the omitted
subject of the infinitive is coreferential with the matrix subject
(as with promise). This interpretation is, for some speakers, more
readily obtainable if the hierarchical relation between matrix subject and object is such that a request for permission is more likely
than a directive:

> (31)   Little Mary asked the teacher to leave the room.

What is interesting is that the possibility of this secondary reading, for at least a wide range of speakers of English, jumps dramatically if the infinitive construction is expanded to include explicit reference to the granting of permission:

> (32)   Little Mary asked the teacher to be allowed to leave the
room.

In fact, with this example, the interpretation where Little Mary is
interpreted as subject of to be allowed to leave the room is dominant, and the other interpretation approaches pragmatic incoherence

('little Mary requested of the teacher that the teacher be allowed to leave the room'). Note that with these examples with <u>ask</u>, the possibility in principle of two readings is guaranteed by the lexical ambiguity of <u>ask</u> ('request' or 'request permission for'). In general, the interpretation 'request' is dominant, with object-control of the omitted subject of the infinitive.[8] Where, however, other parts of the sentence make it pragmatically more reasonable that the interpretation is 'request permission for' (for instance, by referring explicitly to the component of permission), the interpretation 'request permission for' is dominant, with subject-control of the omitted subject of the infinitive.

To summarize the discussion of section 2: Certain constellations of speaker, hearer, and actor are more natural pragmatically, for instance the speaker also being the actor with promises, or the hearer also being the actor with directives. The formally less marked constructions, with omission of the embedded subject, correspond closely to those situationally less marked constellations. Some of the exceptions indicate interaction with formal principles, in particular the minimal distance principle, but several of the idiosyncrasies of matrix verbs in this area find a fuller explanation in terms of interaction with more specific pragmatic principles, in particular the effect of overt specification of other pragmatic parameters.

3. Transitivity and morphology

In recent work on the morphological expression of agents and patients in transitive constructions, stemming in large part from the work of Silverstein (1976), it has been noted that several cross-linguistically valid generalizations can be stated in terms of a hierarchy of salience of noun phrases. This hierarchy consists of at least two subhierarchies, given as (33) below, such that noun phrases to the left of either hierarchy are more salient than those to the right:

(33)   a. Definite   Specific indefinite   Nonspecific

        b. 1st, 2nd person   Other human   Other animal
            Inanimate

As agents in transitive constructions, noun phrases lower in salience are more likely to have overt morphological marking, i.e. to appear in the ergative case. As patients of transitive verbs, noun phrases higher in salience are more likely to have overt morphological marking, i.e. to appear in the accusative case. Examples (34)-(47) illustrate this generalization from a small number of languages, although it should be noted that instantiations of these generalizations are extremely widespread across languages, and indeed represent one of the best established examples of empirical crosslinguistic universals.

In Turkish, transitive patients take no suffix if they are indefinite, but require the suffix -I if definite:

(34)   Ali bir kitap aldI.

        Ali a book bought

        'Ali bought a book'

(35) 'Ali kitab-I aldI.

        Ali book ACC bought

        'Ali bought the book.'

In Yidiny, an Australian Aboriginal language, first and second
person pronouns take no suffix as intransitive subjects or as trans-
sitive agents, but take an accusative suffix as transitive patients.
Conversely, noun phrases lower on the hierarchy (b) take no suffix
as intransitive subjects or as transitive patients, but take the
ergative suffix as transitive agents (Dixon 1977:256):

(36)   Guda:ga gadang.

        dog   come

        'The dog is coming.'

(37)   Waguja-nggu guda:ga bunja:ny.

        man-ERG   dog   hit

        'The man hit the dog.'

(38)   Wagu:ja gudaga-nggu baja:l.

        man   dog-ERG   bit

        'The dog bit the man.'

(39)   Nganyji gadang.

        we   come

        'We are coming.'

(40)   Nganyji nyuni-ny bunjang.

        we   you-ACC   hit

        'We will hit you.'

(41)   Nyundu   nganyji:-ny bunja:ny.

        you   we-ACC   hit

        'You hit us.'

(42)   Nganyji bana   wungang.

        we   water will-drink

        'We will drink water.'

(43)   Bana:-ng nganyji:-ny jaja:l.

        water-ERG we-ACC   not-like

        'The water doesn't like us.'

In Armenian, both hierarchies (a) and (b) are relevant, since the accusative suffix is used only for transitive patients that are both animate and definite:

(44)   Mek mard tesa.

        a man   I-saw

        'I saw a man.'

(45)   Mek girk^h tesa.

        a   book I-saw

        'I saw a book.'

(46)   Girk^h-ə   tesa.

        book-DEF I-saw

        'I saw the book.'

(47)   Mard-u-n   tesa.[9]

        man-ACC-DEF I-saw

        'I saw the man.'

The formal generalization derivable from the above data is that less formal marking correlates with an agent higher in salience and a patient lower in salience. In Southern Tiwa, this same salience differential is seen in the verb morphology rather than in the noun

morphology: where the patient is higher in salience than the agent,
the formally marked passive voice must be used (Allen and Frantz
1978:12):

   (48)   Ti-seuan-mu-ban.

        I/him-man-see-PAST

        'I saw the man.'

   (49)   Seuani-de a-mu-che-ban.

        man-INSTR you-see-PASS-PAST

        'The man saw you.'

We can now proceed to ask how this formal generalization correlates with properties of the world, or rather with people's conceptualization of the real world. Developing the hypothesis set out
in Silverstein (1976), we can claim that for people the most natural
situation is one in which the referent of the agent is relatively
high in salience and the referent of the patient relatively low in
salience. In other words, humans regard as more natural situations
where entities closer to themselves are acting upon entities less
close to themselves, rather than vice versa: closest are other
participants in the speech act (i.e. first and second person pronouns), then other humans, then other animals, and finally inanimate objects. Thus the mass of data presented in this section boils
down to a single simple correlation between markedness of situations
and markedness in morphological structure: morphological unmarkedness corresponds to greater unmarkedness of the situation, while
overt morphological marking corresponds to a greater markedness of
a (less expected) situation.

4. Conjunction reduction

In English, when transitive and intransitive clauses are co-
joined as in (50), it is possible to omit the subject of the second
clause, provided that it is coreferential with the subject of the
first clause:

   (50)   The man hit the doctor and (the man) ran away.

To avoid prejudging the issue of assigning grammatical relations in
languages of different types, a more neutral terminology will be
used, under which both the trigger and the target of conjunction
reduction in English must belong to the set consisting of intransitive subjects and transitive agents. In the Australian Aboriginal
language Dyirbal, a similar constraint exists, except that here
trigger and target must both belong to the set consisting of intran-

sitive subjects and transitive patients (Dixon 1972:
130-134):

(51)   Jugumbil yara-nggu balgan, baninyu.

        woman   man-ERG   hit   came-here

        'The man hit the woman, and the woman came here.'

In both English and Dyirbal, conjunction reduction is constrained by
a purely formal syntactic constraint. This constraint is so strong
that it overrides all other considerations, such as those of real-
world likelihood, with the result that the only interpretation as-
signable grammatically to sentence (52) below is the absurd one
where the speaker splattered:

(52)   I dropped a papaya and splattered.

Although it is far more likely that the papaya splattered as a result
of my having dropped it, this interpretation is not grammatically
available to (52).

In this section, we will be concerned with languages which do
not have a strict constraint on conjunction reduction. In this way,
we will be able to investigate, for these languages, the interrela-
tion between formal lack of markedness (omission of a noun phrase)
and lack of markedness (greater naturalness, expectedness) of corresponding situations. Languages like English and Dyirbal, with strict
syntactic constraints on conjunction reduction, are simply irrelevant
to this exercise, although at the end of this I return to some possible links between languages of both types.

In Kalaw Lagaw Ya, a language spoken in the Torres Strait Islands
between Australia and New Guinea and belonging genetically to the
Australian language family, there is no syntactic constraint on the
interpretation of corresponding sentences, so that sentence (53) has
two possible interpretations:

(53)   Kala   Gibuma-n   mathaman a   zilamiz.

        Kala   Gibuma-ACC   hit-SG   and ran-away-SG

        'Kala hit Gibuma and Kala/Gibuma ran away.'

In order to follow the argument below, it will be necessary to present some material on Kalaw Lagaw Ya morphology; for further details,
reference may be made to Comrie (1981; conjunction reduction is discussed on pages 31-36). Noun phrases have various morphological
systems for marking intransitive subjects, transitive agents, and
transitive patients: proper names have a marked accusative case for

transitive patients; nonplural common nouns have a marked ergative case for transitive agents; singular personal pronouns have a marked ergative for transitive agents and a marked accusative for transitive patients; plural common nouns and nonsingular pronouns use the citation form for all three of these relations. Verbs have three distinct number forms: singular, dual, and plural; intransitive verbs agree in number with their subject, transitive verbs with their patient.

Although sentence (53) above is ambiguous, there is strong preference for the interpretation 'Kala hit Gibuma and Kala ran away', a property to which I return at the end of this section. The fact that both interpretations are allowed by the grammar, however, can be seen by looking at examples where other properties of the sentences force one interpretation or the other. Thus, if the two noun phrases in the transitive clause are of different number, the number agreement on the verb in the intransitive clause will force one interpretation:

(54) Kala yoepkoez-il mathamoeyn a   zilamiz.

      Kala woman-PL   hit-PL   and ran away-SG

      'Kala hit the women and Kala ran away.'

(55) Kala yoepkoez-il mathamoeyn a   zilamemin.

      Kaka woman-PL   hit-PL   and ran-away-PL

      'Kala hit the women and the women ran away.'

In (54), the verb zilamiz is singular, so the only potential antecedent is the preceding transitive agent Kala; in (55), the verb zilamemin is plural, so yoepkoezil 'the women' is the only potential antecedent, i.e. the transitive patient. Similarly, where the clauses appear in the order intransitive before transitive, the target may in principle be either agent or patient of the transitive clause, although other aspects of the morphology may force one or the other interpretation:

(56) Kala ngapa a   Peku-n   mathaman.[10]

      Kala came and Peku-ACC hit-SG

      'Kala came and Kala hit Peku.'

(57) Kala ngapa a   Peku mathaman.

      Kala came and Peku hit-SG

      'Kala came and Peku hit Kala.'

In (56), <u>Pekun</u> is overtly accusative, therefore must be patient, which leaves open only the possibility of interpreting <u>Kala</u> as agent of the transitive clause. In (57), <u>Peku</u> is not marked morphologically, and since it is a proper name it cannot be patient, but must be agent, whence by elimination <u>Kala</u> must be patient. Disambiguation by number in such intransitive--transitive sequences is illustrated in (58)-(59):

(58) Ngoey ngapa a   ngipel mathamoeman.

we-PL came   and you-DU hit-DU

'We came and we hit you two.'

(59) Ngoey ngapa a   ngipel mathamoeyn.

we   came and we-DU hit-PL

'We came and you two hit us'

Although the nonsingular pronouns in (58)-(59) have no morphological marking, the interpretation of each sentence is unequivocal, since the transitive verb must agree in number with its patient.

In the examples discussed so far, all interpretations allowed by the structure of the sentence have been permitted, with some preference in the ambiguous sentence (53) for coreference with the agent rather than the patient of the transitive verb. In the following examples, although the grammatical structure of the sentence leaves two interpreations open, in fact only one interpretation is offered by native speakers, crucially even in sentences like (61) and (63), where the intransitive subject is coreferential with a transitive patient rather than a transitive agent:

(60) Kala woerab   woeriman   a   zilamiz.

Kala coconut struck-SG and ran-away-SG

'Kala struck the coconut and Kala ran away.'

(60) Kala woerab   woeriman   a   papalamiz.

Kala coconut struck-SG and broke-SG

'Kala struck the coconut and the coconut broke.'

(Note that <u>papalamiz</u> in (61) is intransitive 'break'.)

(62) Ngoey ngapa a   woerab-al purthamoeyn.

          we-PL   came   and   coconut-PL   ate-PL

'We came and we ate the coconuts.'

(63)   Woerab-al   paremin a   ngoey purthamoeyn.

       coconut-PL fell-PL and we-PL ate-PL

       'The coconuts fell and we at the coconuts.'

In principle, no doubt, sentences (60)-(63) are ambiguous, having also interpretations '...and the coconut ran away', '...and Kala broke', '...and the coconuts ate us'. The reason why the inter-pretations cited as glosses are in fact assigned is because of the correlation between unmarked situation and unmarked form: the un-marked situation is for people, rather than coconuts, to run away; for coconuts, rather than people to break; and for people to eat coconuts, rather than vice versa. If for some reason one wanted to express the less expected situation, e.g. in narrating fiction, then it would be necessary to make the intended reference explicit in the sentence.

A number of languages have been shown to have systems like that of Kalaw Lagaw Ya, but interestingly these languages differ on some small points, thus providing variations on the Kalaw Lagaw Ya pat-tern. The following data on Lenakel, an Austronesian language of southern Vanuatu, are taken from Lynch (1983). In Lenakel, there is a prefix m- which can be used on noninitial clause verbs to indi-cate that the subject of that verb is the expected subject in terms of coreference with a noun phrase in the preceding clause; Lynch calls this prefix 'echo subject' (ES). Where two intransitive clau-ses are conjoined, m- indicates that the second has the same intran-sitive subject as the first; if the two subjects are noncoreferen-tial, then the regular subject person-and-number prefix must be used on the second verb:

(64)   I-im-vin   m-im-apul.

       I-PAST-go ES-PAST-sleep

       'I went and slept.'

(65)   I-im-vin   r-im-apul.

       I-PAST-go he-PAST-sleep

       'I went and he slept.'

Let us now consider Lenakel examples where a transitive and an intransitive clause are conjoined, in that order, the point of in-terest being whether the echo subject prefix is interpreted as core-ferential with the preceding transitive agent or the preceding transitive patient. Where both interpretations are probable, as in (66), Lenakel has a very strong preference for coreference with the agent:

(66)   Magau r-im-ho   Tom kani m-akimw.

        Magau he-PAST-hit Tom and ES-run-away

        'Magau hit Tom and Magau ran away.'

However, Lynch (personal communication) informs me that, if the context forces the other interpretation, then the interpretation 'Magau hit Tom and Tom ran away' would be possible. If the echo subject prefix in (66) is replaced by the third person subject prefix, then it is totally impossible to interpret Magau as the one who ran away, i.e. this forces an interpretation where transitive agent and intransitive subject are noncoreferential; the one who ran away was either Tom, or some other third person not mentioned in the sentence:

(67)   Magau r-im-ho   tom kani r-akimw.
        Magau he-PAST-hit Tom and he-run-away
        'Magau hit Tom and he (perhaps Tom, but definitely not
        Magau) ran away.'

This differs, incidentally, from the closest parallel in Kalaw Lagaw Ya: if an overt third person pronoun is inserted into the second clause of Kalaw Lagaw Ya sentence (53), it remains ambiguous, just as in English:

(68)   Kala Gibuma-n mathaman a   nuy zilamiz.

        Kala Gibuma-ACC hit-SG   and he   ran-away-SG

        'Kala hit Gibuma and he (Kala/Gibuma) ran away.'

In Lenakel, then, the difference between echo subject and regular subject prefix comes close to forcing an agent = subject versus an agent = subject interpretation: with the echo subject marking, the agent = subject reading is very strongly preferred; with the regular subject prefix, only the agent = subject interpretation is available.

Despite the prevalence of the agent = subject interpretation where the echo subject prefix is used, this can be overridden if other aspects of the sentence indicate that it should be. These other aspects may be formal, as in (69), where the dual verb in the second clause allows only the transitive patient of the first clause as trigger:

(69)   Magau r-im-ho   perasuaas mil kani m-u-akimw.

        Magau he-PAST-hit girl   DU and ES-DU-ran-away
        'Magau hit two girls, and the two girls ran away.'

More importantly in light of our present concern, the agent = subject interpretation is also overridden in examples like (70), where the relevant factor is the greater real-world likelihood of the papaya splattering, rather than of the speaker splattering, as a result of the speaker dropping the papaya:

(70)   I-im-alak-hiaav-in   kesi   m-pwalhepwahle.

I-PAST-throw-down-TRANS papaya ES-splatter

'I dropped a papaya and the papaya spattered.'

The material in the present section can thus be summarized as follows: In addition to the formal principles that may restrict the reference of omitted noun phrases coreferential with preceding noun phrases, many languages also provide evidence for a principle linking formal markedness with situational markedness: the formally unmarked construction, with omission of the noun phrase or expression of the noun phrase by a maximally neutral expression (such as the Lenakel echo subject), is more likely to be used to refer to the unmarked situation, i.e. the situation that is more expected in terms of our conceptualization of the real world.

Before leaving conjunction reduction, however, I would like to make one further set of comments. In the examples treated above where sentences were potentially ambiguous, as in Kalaw Lagaw Ya (53) or Lenakel (66), it was noted that the predominant interpretation is one where the omitted intransitive subject is coreferential with the transitive agent, rather than the transitive patient – this predominance is particularly strong in Lenakel. A similar imbalance of potential interpretations is noted by Nedjalkov (1979:242-243) for Chukchee, a Chukotko-Kamchatkan language of eastern Siberia:

(71)   atlag-e   ekak talayvanen ank?am ekvetg?i.

father-ERG son he-hit-him and   he-left

'The father hit the son and the father/the son left.'

Although both interpretations are possible, the preferred interpretation is that the father left. If, however, ekak 'the son' is topicalized in the first clause by preposing it, then the relative preference shifts, i.e. now it is the son who left:

(72)   Ekak atlag-e   talayvanen ank?am ekvetg?i.

son   father-ERG he-hit-him and   left

'As for the son, the father hit the son and the son left.'

In fact, in our discussion so far of conjunction reduction, we have

omitted one important function of this construction: it enables the
speaker to maintain topic continuity across a sequence of clauses.
The importance of topic continuity explains the difference in preferred readings of Chukchee sentences (71) and (72). It is known
independently that there tends to be a high correlation in language
between agent and topic, therefore one would expect there to be a
bias, in the absence of overt indication that some noun phrase other
than the agent is topic, in favor of taking the agent as topic, and
thus as the noun phrase that triggers and is target of conjunction
reduction. In English this correlation has become rigidly gramma-
ticalized by the syntactic constraint on conjunction reduction.[11]
In Kalaw Lagaw Ya, Lenakel, and Chukchee this correlation can be
seen in the preference for taking the agent as trigger for conjunction reduction, except where (at least in Chukchee) there is overt
indication of some other noun phrase as topic (or, of course, other
aspects of the structure of the sentence or real-world probabilities
suggest that the trigger is a non-agent). This leaves only the rare
Dyirbal type, where the preferred (in Dyirbal, the only) trigger is
the patient of a transitive clause, as a purely formal type not related to any nonformal principle.

5. Conclusions

     In this paper I have tried to show, on the basis of a range of
syntactic and morphological phenomena, that one can often establish
a correlation between linguistic markedness and situational markedness, i.e. that those constructions that involve less formal markedness linguistically correspond to those extralinguistic situations
which - in fact or in our conceptualization of those situations -
are more expected. If this claim can be maintained in general, then
it suggests that linguistic markedness may not be purely a formal
property of language or an accidentally inherited property of human
beings, but rather that it may be explainable in large measure in
terms of human interaction with other humans and with the world
around them.

Footnotes

     *This is a revised version of my presentation to the Symposium
on Markedness at the University of Wisconsin-Milwaukee in March,
1983. I am grateful to all participants who offered questions and
comments on this presentation.

     The following abbreviations are used: ACC (accusative), DEF
(definite), DU (dual), ERG (ergative), ES (echo subject), GEN
(genitive), INSTR (instrumental), LOC (locative), PASS (passive), PL
(plural), SG (singular), TRANS (transitive).

1. For help with the Armenian examples here and below, I am grateful to G. Mastian.

2. The use of the citation form, as in (3), is restricted to the colloquial language (where (4) and (5) are also possible). Minimal pairs like (4) and (5) are sometimes given slightly different interpretations, e.g. 'in Erevan (including suburbs)' versus 'inside Erevan (the city itself)', but this is not consistent: thus both tupʰ-um and tupʰ-i meў mean 'in the box'.

3. Compare the statement by Minassian (1980:301) that type (3) is restricted to 'verbes de repos' (verbs of rest).

4. For a minority of English speakers, sentence (18) is either ungrammatical, or has the interpretation 'Tom promised Sally that Sally would return before nightfall'. This apparently reflects domination of the formal (minimal distance) principle over the pragmatic principle outlined in the text, for these speakers. What is crucial to my argument is that some speakers have the interpretation assigned to (18) in the text, as this demonstrates that the minimal distance principle is not sufficient to account for this range of phenomena in all varieties of English. In fact, the interpretation given for (18) in the text is the one that most English speakers get, and is also at least the prevalent interpretation in translation equivalents in a number of languages.

5. The construction persuade NP that...should has, in addition to the interpretation where NP is exhorted to some course of action, an interpretation where NP is simply persuaded of the truth of some proposition; this latter interpretation is irrelevant for present purposes.

6. Strictly speaking, of course, sentence (24) does not imply that Tom has influence over whether Walter Cronkite interviews him, but rather that Tom is claiming that Tom believes that Tom has influence... For accuracy, similar modifications should be made to the other statements in the text.

7. Presumably, this reflects interaction with the minimal distance, or some similar, principle, as this interpretation parallels that of (26), although the details remain to be worked out.

8. This may either be an arbitrary property of the lexical representation(s), or a reflection of the minimal distance principle.

9. The definite article has the form -ə after a consonant, -n after a vowel; -n is also used, at least in the standard language, after a consonant when the following word begins with a vowel. The suffix we have labeled 'accusative' is called 'genitive' and/or 'dative' in traditional grammar, since it also encodes possessors and recipients;

what is crucial for present purposes is that it marks a transitive
patient high on the hierarchy of salience.

10. Ngapa is a nonverbal intransitive predicate, and does not agree
in number with its subject.

11. In English, the only way round the constraint is to topicalize
the patient via passivization, to give topic continuity between the
two (derived) intransitive subjects, as in the doctor was hit by the
man and (the doctor) ran away.

References

Allen, Barbara J. and Frantz, Donald G. 1978. 'Verb agreement in
    Southern Tiwa'. Proceedings of the Fourth Annual Meeting of the
    Berkeley Linguistics Society, 11-17.
Chomsky, Noam and Halle, Morris. 1968. The sound pattern of
    English. New York: Harper and Row.
Comrie, Bernard. 1981. 'Ergativity and grammatical relations in
    Kalaw Lagaw Ya (Saibai dialect)'. Australian Journal of Linguis-
    tics. 1, 1-42.
Dixon, R.M.W. 1972. The Dyirbal language of North Queensland.
    Cambridge: Cambridge University Press.
Dixon, R.M.W. 1977. A Grammar of Yidiny. Cambridge: Cambridge
    University Press.
Greenberg, Joseph H. 1966. Language universals, with special
    reference to feature hierarchies. The Hague: Mouton.
Lynch, John. 1983. 'Switch Reference in Lenakel'. In Haiman, John
    and Pamela Munro, eds., Switch reference and universal grammar.
    Amsterdam: Benjamins.
Minassian, Martiros. 1980. Grammaire d'arménien oriental.
    Delmar, NY: Caravan Books.
Nedjalkov, V.P. 1979. 'Degrees of ergativity in Chukchee'. In
    Plank, F., ed., Ergativity: towards a theory of grammatical re-
    lations. London: Academic Press. Pages 241-262.
Postal, Paul M. 1970. 'On coreferential complement subject
    deletion'. Linguistic Inquiry 1, 439-500.
Rosenbaum, Peter S. 1967. The grammar of English predicate
    complement constructions. Cambridge, MA: MIT Press.
Searle, John R. 1969. Speech acts: an essay in the philosophy of
    language. Cambridge: Cambridge University Press.
Silverstein, Michael. 1976. 'Hierarchies of features and
    ergativity'. In Dixon, R.M.W., ed., Grammatical categories in
    Australian languages. Canberra: Australian Institute of
    Aboriginal Studies. Pages 112-171.

MARKEDNESS AND DISTRIBUTION IN PHONOLOGY AND SYNTAX

Jeanette K. Gundel, Kathleen Houlihan, and
Gerald A. Sanders

University of Minnesota

1. Introduction

The notion of markedness has been used, in a variety of distinct
but possibly related senses, for the description and analysis of phenomena involving both the syntactic and phonological structures of
human languages. It has frequently been assumed that there is a
single and relatively clear sense of markedness that is equally
relevant to phonology and syntax alike, and this assumption, in fact,
constitutes a crucial precondition for a number of general linguistic hypotheses that have recently been proposed--for example, Eckman
(1977) and Comrie (1984). The basic notion of markedness, however,
and a number of the most fundamental concepts associated with it--
concepts such as "neutralization", "mark", "opposition", "privative",
and "bilateral"--were first developed and exemplified, by Trubetzkoy
(1939[1969]) and others, on the basis of phonological data alone.
It thus remains to be determined whether this notion is indeed applicable in the domain of syntax or not. This paper will seek to
address this question and to clarify thereby the general notion of
markedness itself and its significance in the development of
linguistic hypotheses.

The primary departure point for our investigation will be the
central notions of typological markedness and contextual distribution. Typological markedness is an asymmetric binary relation between linguistic forms in the domain of all human languages, such
that the presence of one form in a language implies the presence of
the other but not the reverse. This is stated as a definition in (1).

TYPOLOGICAL MARKEDNESS (definition):
A is typologically marked relative to B (and B is typologically
unmarked relative to A) if and only if every language that has A
also has B but not every language that has B also has A.

Typological markedness relations thus hold within the domain of
all languages, independently of context or attendant circumstances.
Contextual distribution relations—such as free variation, contrast,
defective distribution and neutralization—hold, on the other hand,
within a particular language and with respect to particular contexts
in the expressions of that language. Distributional relations and
typological markedness are therefore logically and methodologically
distinct and wholly independent of each other.

Given their logical independence, then, there would be no
a priori reason to expect any systematic relationship to hold between markedness and distributional relations in natural languages.
Yet it does appear to be the case that some such relationship
obtains, namely, that it is the typologically unmarked member of a
markedness relation that occurs in positions of neutralization or
defective distribution in a given language. This general correlation, stated as an absolute, is given in (2).

(2) Typological markedness and difference of distribution are
    correlated such that, in a given language with two alternating
    forms A and B, if A has a wider distribution than B, then A is
    typologically unmarked relative to B, and if A is typologically
    unmarked relative to B, then A has a wider distribution than
    B.[1]

The idea of there being a link between some type of markedness
and some types of distributional relations goes back to Trubetzkoy
(1939[1969]), of course, who essentially determined markedness, in a
language-specific sense, in terms of distributional relations within
a single language. With respect to markedness in the universal
sense, an apparent correlation between markedness relations and
certain distributional phenomena has been noted and discussed
generally, for example, by Greenberg (1966), and with particular
reference to the nature and function of phonological rules, by
Houlihan and Iverson (1979, 1980).

However, the precise nature and scope of the correlation has not
yet been systematically investigated; nor has the full range of distribution relations that obtain been considered, either in phonology
or, especially, in syntax. It is our primary purpose in the present
paper to undertake this task of systematically investigating the
relationship between distribution and markedness in both phonology
and syntax, as a basis for determining the possible place of markedness in the both domains and its significance for linguistic inquiry
in general.

2. Distribution and Markedness

The significant structural characteristics of linguistic expressions are a function of their distribution and interpretation. The
interaction between distribution and interpretation yields three
simplex relations schematized in (3)--contrast (3a), free variation
(3b), and defective distribution (3c).

(3) Simplex Distributional Relations

    (a) Contrast: (b) Free Variation: (c) Defective Distribution
       A/X   A/X   A/X
       B/X   B/X   *B/X
       A<A   A<[A,B]
       B<B   B<[A,B]

(where "/" signifies "in the context of", "<" signifies " is
interpreted as", and "[P,Q]" signifies "P or Q"[2].)

These three simple distributional relations can be combined and
elaborated in various ways in natural languages to yield a variety
of more complex ones.[3] We will now examine in turn each one of
these complex distributional relations in phonology and syntax, with
the purpose of determining the precise nature of the relationship
between distribution and typological markedness in each particular
case.

3. Complex Distributional Relations

3.1 Simple defective distribution with contrast

The first type of complex distributional relation that we will
examine is of a type that involves contrast in one environment and
defective distribution in another. Here two forms A and B both
occur in some environment X with different interpretations, and
hence are in contrast, while only A occurs in some other environment
Y, where it is interpreted only as A. This particular type is
schematized in (4).

(4) Simple defective distribution with contrast:[4]

    A/X   A/Y
    B/X   *B/Y
    A<A   A<A
    B<B

3.1.1 Phonology

A phonological example of this type of defective distribution is
provided by the distribution of voiced and voiceless obstruents in
English. These contrast in most environments, but there is no con-

trast after syllable-initial /s/, since voiced obstruents do not occur in this position.[5]

The typological facts regarding voiced and voiceless obstruents are that there are languages like English with both, languages like Hawaiian with only voiceless obstruents, but no language with only voiced obstruents.[6] It follows, then, that voiceless obstruents are typologically unmarked with respect to voiced ones. Therefore, the member of the opposition found in the position of defective distribution in English, the voiceless obstruent, is the unmarked member, which is consistent with the correlation between markedness and distribution stated in (2). In other words, the unmarked member has a wider range of distribution than the marked. Other phonological examples in this category all appear to observe the same correlation. Such examples include the defective distribution of marked /ŋ/ vs. unmarked /n/ word-initially in English and the defective distribution of marked aspirated and voiced obstruents vs. unmarked voiceless unaspirated obstruents word-finally in Thai. In all these cases, the member of the opposition that has the wider distribution (i.e., which is not defectively distributed) is the unmarked member, which is consistent with the correlation between markedness and distribution in (2).

3.1.2 Syntax

A syntactic example of defective distribution with contrast is provided by the ordering of subjects and modals in English. Thus, as illustrated in (5), the two orderings are in contrast in main clauses, where subject before modal has a declarative interpretation, as in (5a), and modal before subject has an interrogative interpretation, as in (5b). In subordinate clauses, however, as shown in (5c) and (5d), the contrast is suspended in favor of the subject-modal order.

(5) a. Mary can swim.
    b. Can Mary swim?

    c. I know ⎡whether⎤ Mary can swim.
               ⎢if   ⎥
               ⎢how   ⎥
               ⎣that   ⎦

    d. *I know ⎡whether⎤ can Mary swim.
                ⎢if   ⎥
                ⎢how   ⎥
                ⎣that   ⎦

There appears to be a clear typological markedness relation here. Thus it seems that every language with sentences of the form Modal-Subject-X, for example English and German, also have sentences of

the form subject-modal-X. However, there are some languages, such
as Thai, that have Subject-Modal-X but not Modal-Subject-X. Thus,
as in phonology, it is the unmarked member of the markedness
relation that occurs in the wider range of contexts here and the
marked member that is defective in its distribution.

Another example of defective distribution with contrast is provided by relativization in English. In this language, relativization is generally not restricted to any single NP position in the
relative clause.   Thus, for example, a subject can be relativized,
as in (6a), or an object can be relativized as in (6b).

(6) a.   The woman who [ø saw the burglar] is here.

    b.   The woman who [the burglar saw ø] is here.

However, in complements introduced by that or some other overt
complementizer, only non-subjects can be relativized, as illustrated in (7).

(7) a.   *The woman who the police reported that
          [ø saw the burglar] is here.

    b.   The woman who the police reported that
         [the burglar saw ø] is here.

There appears to be a typological markedness relation here. According to Keenan and Comrie (1977), all languages allow relativization
of subjects. Moreover, there are some languages, e.g. Tagalog,
which allow relativization only of subjects. Thus, relativization
of non-subjects implies relativization of subjects; but relativization of subjects does not imply relativization of non-subjects.
Subject relativization is therefore typologically unmarked with
respect to relativization of non-subjects. In this case, then,
unlike the previous one (and also unlike the phonological examples
in this category) it would be the unmarked member of a typological
markedness relation that is defective in its distribution and the
marked member that has wider distribution. (We will return to this
apparently anomalous case in section 4.)

3.2   Simple Defective Distribution with Free Variation

The second category to be considered is the case of simple
defective distribution with free variation, schematized in (8).

(8) Simple defective distribution with free variation

    A/X   A/Y
    B/X   *B/Y
    A < [A,B]   A < A
    B < [A,B]

Here both A and B occur in some environment X, but their interpretations there are not distinct; and only A occurs in another environment Y, interpreted there only as A. In other words, A and B are in
free variation in X, and B is defectively distributed relative to A
in Y.

3.2.1   Phonology

A phonological example of this category is provided by the distribution of released and unreleased stops in English. Both occur
in free variation in word-final position, but only released stops
occur word-initially.

The typological facts regarding released and unreleased stops
are that there are languages like English that have both, languages
like Fijian that have only released stops (since they have only open
syllables), but no languages that have only unreleased stops. From
this it follows that released stops are typologically unmarked with
respect to unreleased ones. Thus both marked and unmarked stops
occur word-finally in English in the position of free variation,
while the released stops that occur word-initially in the position
of defective distribution are unmarked. Here, as in all known phonological cases of this kind, both the marked and unmarked forms occur
in the position of free variation, but only the unmarked ones occur
in the position of defective distribution.

3.2.2   Syntax

An example from syntax of defective distribution in one environment with free variation in another is provided by the alternation
in English between dislocational, or double argument, sentences like
(9a) and their non-dislocational paraphrases, as in (9b).

(9) a. The steaks, I'm going to cook them on the grill
    b. I'm going to cook the steaks on the grill

Alternants of this sort are in free variation in main clauses. In
certain embedded contexts, however, as shown in (10), only the non-
dislocational ones occur.

(10)   A. *The grill that the steaks I'm going to cook them on is
           in the garage
       B. The grill that I'm going to cook the steaks on is in the
           garage

Dislocational sentences, which can be schematically represented
as in (11a), are apparently found in all human languages. On the
other hand, if we restrict our domain to NP's functioning as terms—
that is to subjects, objects and other primary arguments of simple
predicates—then it is not the case that every language has sentences

of the non-dislocational and anaphora-free type schematized in (11b).

(11) a. NPi X PROi Y
      b. X NP Y

(where X and Y do not contain pronouns or other anaphoric
constituents coreferential with NP)

Thus, the latter sentence type will be absent, in particular, in any
language that has obligatory agreement between the verb and one of
its arguments in such pronominal features as person, number, and
gender. There are, of course, many languages of this sort, including, for example, Amharic, Arabic, Basque, Fijian, Ojibwe, and Zulu.
With respect to NPs that are terms, therefore, there are languages
that have sentences of the form (11a), but no sentences of the form
(11b), but there are no languages with (11b) but not (11a). Here
then we have a markedness relation such that structures of the form
(11a) are unmarked with respect to structures of the form (11b).
The free variation between these two structures in English would
thus be an alternation between the marked and unmarked terms of a
typological markedness relation that is suspended in certain subordinate clause contexts in favor of the typologically marked term,
contrary to the general correlation in (2).

Another example of free variation between syntactic structures
consists of the alternation in English between structures of the
form Verb-Particle-NP, as in (12a), and structures of the form
Verb-NP-Particle, as in (12b).

(12) a. I let out the cat
      b. I let the cat out

There are certain contexts, however, namely, in certain nominalizations of the type exemplified in (13), where the alternation is suspended in favor of the Verb-Particle-NP structure.[8]

(13) a. My letting out of the cat surprised everyone
      b. *My letting of the cat out surprised everyone

Although it might be expected, perhaps, that the variant that
occurs in the position of suspended alternation here, Verb-Particle
-NP, would be typologically unmarked relative to the variant that is
defectively distributed, Verb-NP-Particle, this is in fact not the
case. Thus, although there are some languages that have Verb-
Particle-NP but not Verb-NP-Particle, such as French, for example,
as shown in (14), there are also other languages, such as German, as
shown in (15), which have Verb-NP-Particle but not Verb-Particle-NP.

(14) a. Il abuse de son autorité   'He takes advantage of his
          *Il abuse son autorité de   authority'

b. Elle consent à la règle   'She consents to the rule'
    *Elle consent la règle à

(15) a. *Er macht auf die Tür
          Er macht die Tür auf   'He opens the door'

      b. *Sie holen her den Mann
          Sie holen den Mann her   'They fetch the man'

Thus there is no typological markedness relation between the two
structures at issue here.

There is an alternative analysis of these structures, however,
under which particles are viewed as non-distinct from adpositions
(prepositions and postpositions), and hence where the alternants
would be of the forms schematized in (16).

(16) a. X V Adp NP
      b. X V NP Adp

Under this analysis, it would be the case that there is a markedness
relation between the two structures, with (16a) being the unmarked
member and (16b) the marked. Thus, though there are languages (e.g.
Korean, Japanese) in which adpositions are always postpositional,
these languages do not have structures where either an adposition or
an NP follows a verb. Every language that does permit this either
has both of the structures at issue (as in English and German) or
only the prepositional one (as in French). The prepositional
structure would thus be unmarked after verbs and the postpositional
one marked. Under this analysis, then, the variant that occurs in
the position of suspended alternation would be the unmarked member
of a markedness relation, as is the case with the parallel situation
in phonology.

3.3   Neutralization by Dominance

The next category, schematized in (17), is neutralization by
dominance, or classic neutralization. This type of distribution
involves two forms A and B in contrast in one environment X while
only one of the two forms occurs in another environment Y. The form
that occurs in Y, moreover, can be ambiguously interpreted there as
either A or B.

(17)   Neutralization by dominance

        A/X   A/Y
        B/X   *B/X
        A<A   A< [A,B]
        B<B

3.3.1   Phonology

The standard example of classic neutralization in phonology is
the so-called final devoicing of voiced obstruents in German,
Russian, Sanskrit, and many other languages. In these languages,
voiced and voiceless obstruents contrast in non-final position but
the contrast is neutralized in final position in favor of the voiceless obstruents.

As shown above, voiceless obstruents are typologically unmarked
with respect to voiced ones. Thus the member of the opposition
found in the position of neutralization in these languages is the
typologically unmarked member, which is consistent with the general
correlation between unmarkedness and wider distribution.

Other familiar cases in this category include final neutralization of the aspiration contrast in Sanskrit (Whitney 1889)and the
neutralization of /t/, /s/, and /č/ to [t] in Korean (Kim-Renaud
1974). In each case, contrasts are maintained in some positions but
neutralized to the typologically unmarked member of the opposition
in others.

3.3.2   Syntax

An example of neutralization by dominance in syntax is provided
by NP be NP constructions in English where one of the two NPs is
questioned. The relevant data are presented in (18).

(18)   Echo Questions
        a. Caesar was who? (cf. Caesar was a Roman emperor.)
        b. Who was Caesar? (cf. John was Caesar (in the play).)

(19)   Ordinary Questions
        a. *Caesar was who?
        b. Who was Caesar?

In questions with strong presuppositions (where the wh-form is not
fronted), such as echo questions and questions used in police interrogation, both (18a) and (18b) occur, with the former understood as
a question about the predicate and the latter understood as a question about the subject. In ordinary questions, however, only (19b)
occurs, but with both of these interpretations.

If the relevant terms here are clauses with preverbal question
phrase and clauses with postverbal question phrase, then it again
appears to be the unmarked member which appears in the position of
neutralization here. There are languages like English and Fijian
which have both structures and strict verb final languages like
Japanese which have preverbal question phrases but no postverbal
ones. However, there are no known languages where a question phrase
can occur after verbs but never before them.[9]

So here, as in classic neutralization in phonology, it is the unmarked member of a markedness relation which occurs in the position of neutralization.

Another example of this type of neutralization in syntax involves the distribution of genitive NPs in so-called picture noun phrases in English.

(20)   a. A picture of John's is hanging on the wall
       b. A picture of John is hanging on the wall
       c. John's picture is hanging on the wall
       d. *John picture is hanging on the wall

As shown in (20), both the inflected NP and the non-inflected NP can occur after the preposition <u>of</u>, but with different interpretations. The phrase with the inflected NP in (20a) can only be interpreted as a picture which belongs to John, and the one with the non-inflected NP in (20b) can only be interpreted as a picture which was taken, painted, etc. of John. However, the phrase with the inflected NP before the picture noun, as in (20c), can have either interpretation, and the phrase with the non-inflected NP before the picture noun, as in (20d), is ungrammatical.

With respect to markedness relations, the relevant terms to be compared here are morphologically marked genitive vs. non-morphologically marked genitive. There are languages like English in which a genitive is always morphologically marked, either by preposition or by inflection or by both, and languages like Chinese where the morphological marking of genitives is optional in some cases,[10] but as far as we know, there are no languages in which there is never any morphological marking of genitives. Thus, there is a markedness relation here which fits the correlation. That is, it is the unmarked member of a typological markedness relation which occurs in the position of neutralization.

3.4   Neutralization by Compromise

The last type of defective distribution in one environment is neutralization by compromise, schematized in (21).

(21)   Neutralization by compromise
       A/X   *A/Y
       B/X   *B/Y
       A<A   C/Y
       B<B   C<[A,B]

In this situation, two forms A and B contrast in one environment X, but neither occurs in another environment Y, where a third form C occurs instead that is interpreted as either A or B.

3.4.1 Phonology

A phonological example of neutralization by compromise is provided by the distribution of apical stops and taps in English. The apical stops /t/ and /d/ contrast in most positions in English, but in medial position after a stressed vowel, the contrast can be suspended in favor of an apical tap. (For example, waiting and wading can both pronounced as [wejɾIŋ].)

The typology of apical stops and taps is that there are languages like English that have both stops and taps, languages like Finnish that have stops but no taps, but no languages that have taps without also having stops. Thus, taps are marked with respect to apical stops, and the representative of the /t/-/d/ opposition found in the position of neutralization in English is a marked segment.

This result is different from that found in the case of neutralization by dominance, where the unmarked member of the opposition is found in the position of neutralization. Notice, however, that in neutralization by compromise, the opposition does not neutralize to the marked member of the opposition, but rather to some third form, which just happens to be marked with respect to the members of the opposition.[11] An important fact about all types of neutralization, then, by compromise as well as by dominance, is that the marked member of the opposition is never found in the position of neutralization and therefore does not have wider distribution than the unmarked member.

3.4.2 Syntax

A syntactic example of neutralization by compromise is provided by English independent and relative clauses with non-prepositional datives, as illustrated in (22).

(22) a. I offered the alligator the crocodile
                          IO   DO
      b. I offered the crocodile the alligator
                          IO   DO

      c. (This is) the crocodile I offered the alligator
          =the crocodile I offered the alligator Ø) = (22a)
                                          IO   DO
          =the crocodile I offered Ø the alligator) = (22b)
                                   IO   DO

Thus, (22a) and (22b) contrast in English, but in certain contexts, as in (22c), the clause the crocodile I offered the alligator occurs instead, and it can be interpreted there as either (22a) or (22b).

It is clear that no markedness relationship can hold between
sentences like (22a) and (22b), since they have exactly the same
structure and differ only as to which lexical items occur in which
NP. There does appear to be a markedness relationship, however,
between this structure, i.e. NP V NP NP, and the distinct structure
NP NP V NP, exemplified in (22c). The typological facts are that
there are languages like English which have both structures and
there are languages like Tagalog and German which have NP V NP NP
but not NP NP V NP. As we know of no languages, however, which have
the latter structure but not the former, it appears that NP V NP NP
is unmarked relative to NP NP V NP. Thus, as was the case with the
phonological examples, neutralization by compromise yields a typologically marked form here.

Another example of this type of relationship in syntax involves
ellipsis of a verb and one of its arguments in English subordinate
clauses. Thus, the full clauses Martha likes Mary in (23a) and John
likes Martha in (23b) are in contrast. However, in the context of a
subordinate clause, this contrast is optionally suspended, as in
(23c), in favor of a structure which is formally identical to neither
of the two contrasting structures but which can be interpreted as
either one of them.

(23)   a.   (John likes Mary more than) Martha likes Mary.

         b.   (John likes Mary more than) John likes Martha.

         c.   (John likes Mary more than) Martha   (=(23a) or (23b))

It is clear that all languages allow full clauses, and it is equally
clear that there are languages like English which allow both full
clauses and clauses which consist of only an NP. Since there are
also languages like Chinese which evidently do not allow the latter
structure,[12] a clause consisting of only an NP is typologically
marked relative to a full clause. It is the typologically marked
member, then, which occurs in a position of neutralization here.

Thus, in syntactic cases, as in phonological ones, neutralization
to a third form is found to be correlated with a typological markedness relation such that the third form is typologically marked relative to the forms whose differences it neutralizes. However, since
the form that occurs in the position of neutralization is not the
marked member of the opposition, it does not have a wider distribution than the unmarked form.

3.5 Complementary Distribution

We now turn to examples of defective distribution in both environments, that is, complementary distribution, schematized in (24).

(24)   Complementary distribution
         A/X   *A/Y
        *B/X   B/Y

         A/X=B/Y

Here, there is no contrast between the two forms A and B in either
environment. Rather, both have defective distribution in that only
one occurs in one environment X and only the other occurs in another
environment Y, and they have the same interpretation.

3.5.1 Phonology

   A phonological example of complementary distribution involves
the mutually defective distribution of voiced and voiceless lax
stops in Korean. Voiced lax stops occur only between voiced sounds,
while voiceless lax stops occur only initially, finally, and medially
when adjacent to a voiceless sound (see, e.g., Kim-Renaud 1974).

   It was noted earlier that voiceless stops are typologically unmarked with respect to voiced stops. What we find here, then, is
that the complementary forms are in a markedness relation with
respect to each other and also that the marked as well as the unmarked member of the relation occurs in a position of defective distribution. According to the general correlation between markedness
and distribution, as stated in (2), we would however, expect the
unmarked member to have wider distribution. And that seems to be
the case in this example. Voiced lax stops, the marked forms, only
occur in voiced environments, while the unmarked voiceless stops
occur in all others. There are many other examples in phonology
that exhibit the same pattern, for example the occurence in English,
Spanish and many other languages of (marked) nasal vowels only
before nasal consonants and (unmarked) oral vowels everywhere else.

3.5.2   Syntax

   Complementary distribution of syntactic structures is exemplified by the relation that holds in German between SVO and OVS structures and their corresponding SOV counterparts. Thus, as illustrated in (25), SVO and OVS occur in main clause contexts but not in
subordinate ones, and SOV occurs in subordinate contexts but not as
a main clause.

(25)   a.   Ich kaufte das Haus   (SVO)   'I bought the house'
            Das Haus kaufte ich   (OVS)
           *Ich das Haus Kaufte   *(SOV)

       b. *Ich war dort, als ich Kaufte das Haus   *(SVO)   'I was there
          *Ich war dort, als das Haus Kaufte ich   *(OVS)   when I bought
           Ich war dort, als ich das Haus Kaufte   (SOV)   the house'

SVO and OVS, which are in free variation in main clauses, are thus
both in complementary distribution with SOV, which only occurs in
subordination contexts.

With respect to the complementarity relation between OVS and
SOV, there is a typological markedness relation between the two
structures, with OVS being the marked member and SOV the unmarked.
Every language that has subject before predicate, for example
German, Hixkaryana (Derbyshire 1977) and Russian, also has SOV
clauses, while there are some languages, like English, Japanese and
Korean, that have subject before predicate, but do not have predicate
before subject.

In the case of SVO and SOV, however, both of which have subject
before predicate, there is no typological markedness relation at
all. Thus there are some languages (like German) that have both
structures, others (like English) that have SVO but not SOV, and
still others (like Japanese) that have SOV but not SVO. For this
pair of structures, then, there is a complementarity relation without any associated relation of relative markedness.

Thus neither of the two complementary distribution relations
involved here is analogous to the usual situation for complementary
distribution in phonology. In the case of SOV and OVS, the marked
form, OVS, has a wider distribution than the unmarked form, SOV,
assuming that subordination is a more restricted context than superordination.[13] And, in the case of SOV and SVO, there is no markedness relation at all. (We return to this case in section 4.)

Another example of complementary distribution between syntactic
forms involves the alternation in English between finite verbs,
which occur in tensed non-periphrastic verbal contexts, and
non-finite verbs, which occur in all other verbal contexts. This is
illustrated in (26) and (27).

(26)   a. The cat chases the dog
        b. *The cat chase the dog

(27)   a. *The cat will chases the dog
        b. The cat will chase the dog

Typologically, then, if finiteness in verbs is taken to mean some
limitation, signified by the form of the verb itself, on the range
of individual argument types, tenses, aspects, or adverbials that it
can be associated with in predications, then it is clear that some
languages, such as Thai and Mandarin Chinese, have no finite verbs
at all. There are, on the other hand, no languages, it seems, in
which all verbs are finite.[14] Non-finite verbs are thus unmarked
relative to finite ones.

In English, then, it is the unmarked member of this typological

markedness relation, the non-finite verb, which occurs in the wider
range of environments--in compounds, as modifiers, after modals or
to, etc.--and the marked member which occurs in the narrower range--
as the first verb in a tensed clause. Thus in this case, as in the
standard cases of complementary distribution in phonology, the complementarity relation holds between the marked and unmarked members
of a typological markedness relation, with the unmarked member having
a wider distribution than the marked.

3.6   Distribution in Multiple Contexts

     So far, we have restricted ourselves to a consideration of only
two environments. However, relevant distributional patterns may
often involve more than just two environments. One type involving
three environments is schematized in (28).

(28)   Defective distribution with contrast and free
       variation

       A/X   A/Y   A/Z
       B/X   B/Y   *B/Z
       A<A   A<[A,B]   A<A
       B<B   B<[A,B]

     This pattern involves contrast in one environment, free varia-
tion in a second, and defective distribution in a third. We know of
no syntactic examples of this type of situation, but a phonological
example is provided by the distribution of the tap and trill in
Spanish, illustrated in (29). These two sounds contrast in intervocalic position, shown in (29a), they are in free variation in syllable-final position, (29b), and in word initial position, (29c) (or
after /n/ or /l/), the tap, is defectively distributed relative to
the trill. (Quilis and Fernandez 1966:116)

(29)   Distribution of taps and trills in Spanish

   a. Contrast Intervocalically

      ca[ɾ]o   'expensive'   vs.   ca[r]o   'car'
      pe[ɾ]o   'but'   vs.   pe[r]o   'dog'

   b. Free Variation Syllable-Finally

      pue[ɾ]ta ~ pue[r]ta   'door'
      muje[ɾ] ~ muje[r]   'woman'

   c. Defective Distribution Word-Initially

      [r]osa   (*[ɾ]osa)   'rose'
      [r]io   (*[ɾ]io)   'river'

Thus, the contrast between the alveloar tap and trill in Spanish is not maintained in syllable-final position, where both members of the opposition occur in free variation, or in word-initial position, where only the trill occurs. In a situation of this type, we might expect that the trill should be unmarked with respect to the tap, since it occurs word-initially in the defective distribution environment, and therefore appears to have a wider distribution. However, in this particular case, there is no implicational universal regarding taps and trills in languages of the world. Rather, there are languages like Spanish with both, languages like Irish (De Burca 1970) with only taps, and languages like Finnish (Lehtinen 1963) with only trills. Therefore, taps and trills are not in a markedness relationship with respect to each other. It might appear, then, as in the German word order case, that we have another situation where there is a difference in distribution without a corresponding markedness relation. However, it is not entirely clear that there actually is a substantial difference of distribution between taps and trills in Spanish, since there are additional contexts where only the tap can occur.[15]

A second example of multiple contexts is schematized in (30). It involves contrast in one environment, defective distribution of one of the members of the opposition in a second environment, and defective distribution of the other member in a third.

(30)   Double defective distribution with contrast
       A/X   A/Y   *A/Z
       B/X   *B/Y   B/Z
       A>A
       B>B

A phonological example of this situation can be seen in the distribution of /s/ and /z/ in German, illustrated in (31). These two fricatives contrast in intervocalic position morpheme-internally, shown in (31a), but in syllable-final position, shown in (31b), /s/ but not /z/ occurs (neutralization by dominance), and in stem-initial position before a vowel, (31c), /z/ but not /s/ occurs (simple defective distribution).[16]

(31)   Distribution of /s/ and /z/ in German

       a.   Intervocalically within Morphemes

            wei[s]e   'white'
            wei[z]e   'wise'

       b.   Neutralization Word-Finally

            ein hei[s]es   'a hot one'   hei[s]   'hot
            des Ei[z]es   'of the ice'   Ei[s] (*ei[z])   'ice'

c. Defective Distribution Stem-Initially[17]

| [z]agen | (*[s]agen) | 'say (infinitive)' |
| ge[z]agt | (*ge[s]agt) | 'say (participle)' |

Given the implicational universals regarding voicing of obstruents, /s/ is typologically unmarked with respect to /z/. Although
both members of the opposition have some limits on their distribution, /s/ occurs both syllable-initially and syllable-finally, while
/z/ occurs only syllable-initially. On the basis of position in the
syllable, then, the unmarked member of the opposition has a wider
distribution than the marked.

A similar example in syntax is provided by the distribution of
reflexive and non-reflexive object pronouns in English, as illustrated in (32). The pronouns contrast after verbs like <u>hurt</u>, shown
in (32a), but only the reflexive pronoun occurs with verbs like
<u>behave</u>, (32b), and only the non-reflexive pronouns occurs in clauses
that do not contain an antecedent NP that agrees with the pronoun in
person, number, and gender, (32c).

(32) Distribution of Reflexive and Non-Reflexive Object Pronouns
      in English

      a. <u>Contrast</u>

          John hurt him
          John hurt himself

      b. <u>Defective Distribution of Non-Reflexives</u>

          *John behaved him
          John behaved himself

      c. <u>Defective Distribution of Reflexives</u>

          Mary knew him
          *Mary knew himself

Reflexive pronouns are typologically marked with respect to non-
reflexive pronouns, since every language that has the former also
has the latter, but not vice versa. That is, there are languages
like English that have both types of pronouns, languages like Temne
and Old English that have non-reflexive pronouns but no distinct
pronouns with purely reflexive function, but no language with only
reflexive pronouns. As in the phonological example, both forms are
defectively distributed here, with the unmarked form occurring in

one position of neutralization and the marked form in the other. Nevertheless, although both forms are defectively distributed, the unmarked form seems to occur in a wider range of contexts than the marked. For example, the non-reflexive pronoun can occur in many additional contexts, such as subject position, where the reflexive pronoun cannot occur.

We have presented these examples to illustrate that distributional facts may involve more than two relevant environments and therefore may be considerably more complex than the discussion of instances involving only two relevant environments might indicate. It is clear, therefore, that complex distributional relations of the type just presented must be examined in detail in order to establish precisely the complete set of principles that govern the general correlation between markedness and distribution. So far, though, it still appears that unmarked forms always have wider distribution than the corresponding marked forms, regardless of the number of contexts involved.

4. Discussion and Conclusions

In this paper we have examined a wide range of distributional relations in both phonology and syntax and their interaction with typological markedness relations. We have seen in general in the domain of syntax the same kinds of distribution relations between forms that obtain in the domain of phonology -- contrast, free variation, defective distribution and neutralization by dominance and by compromise. Where the relevant typological facts are relatively clear, moreover, we have found situations in both phonology and syntax where there is a correlation between distributional patterns and typological markedness relations such that the unmarked forms have wider distribution than the marked forms.

It is appropriate now to raise the question of why there should be such a strong correlation between difference of distribution of forms in a single languages and the typological markedness relations of these forms in all languages.

The most reasonable explanation, it seems, is that the correlation is a consequence of the greater communicative effectiveness of the typologically unmarked forms, for example, with respect to language production, comprehension, or learning. Both the typological relation and the distributional relation would follow from this difference in communicative effectiveness, given the additional assumption that human languages develop and maintain themselves in such a way as to resort to the use of a less effective device for a particular purpose only if they have already made use of a more effective one.

However, in spite of the general similarities observed between phonology and syntax here, we have also found in certain cases an

evident lack of parallelism between apparently comparable situations
in the two domains. In particular, we found in three of the eleven
syntactic cases we examined that the typologically unmarked form did
not have wider distribution than the corresponding marked form.
Given these apparently anomalous situations, one possibility, of
course, would be simply to conclude that the relation between markedness and distribution in syntax is not a systematic one at all, or
at least that it is not really comparable to the relationship that
holds in phonology. Before drawing such a conclusion, however,
which ignores the many instances of correspondence between the two
domains, it would seem reasonable first to try to see whether there
might be some other factors at work here -- factors which are wholly
independent of the general relationship between markedness and distribution but which might nevertheless interact with it in such a
way as to override or modify it in some well defined class of situations.

     Although we cannot explore this avenue of inquiry adequately at
the present time, there is at least one independent factor of the
sort we have in mind that may be of particular relevance here with
respect to the three specific cases of syntactic distributional anomaly at issue. Thus, it will be observed that in all three of these
cases -- the alternation of relativized subjects and non-subjects in
English, the alternation between verb-medial and verb-final clauses
in German and the alternation between dislocational and non-dislocational clauses in English -- the crucial difference in environment
is between main clause context, where an alternation occurs, and subordinate clause context, where an alternation is suspended.

     It is worthy of note that this distinction -- between main and
subordinate structures -- has no real counterpart in phonology,
where the relevant distinguishing semantic and pragmatic factors are
inoperative. In syntax, on the other hand, the distinction between
superordination and subordination is one of the fundamental and most
widely operative of all grammatical distinctions. It is also true,
moreover, that variation or diversity of structure is never greater
in the subordinate clauses of a language than in the main clauses.
It is plausible, therefore, that whatever underlies such facts as
these, which are uniquely relevant to syntax, might be operative too
in the three specific cases of syntactic distributional anomaly we
have identified here, since these all involve a distinction between
main and subordinate clauses.

     The importance of the superordinate-subordinate distinction in
syntax and its specific effect on the distribution of syntactic variants in natural languages seems to be due to two independent but
related factors. The first derives from the fact that subordination
may give rise to perceptually complex and potentially misleading
structures which do not arise in the case of simplex clauses, creating special needs, therefore, for compensatory devices for simplification and disambiguation of structures. The second factor stems

from the fact that main and subordinate clauses are generally distinct in the semantic and pragmatic functions that they serve, and hence that the formal characteristics that are suitable to one type of clause may in some instances be incompatible with the functions of the other. Both of these factors appear to be relevant to the cases at hand.

Considering the relativization case first, it will be recalled that within subordinate clauses introduced by an overt complementizer, English relativization appears to be restricted to non-subjects, whereas elsewhere it is permitted for subjects and non-subjects alike. The illustrative data, which was given in (6) and (7), is repeated here for convenience in (33) and (34).

(33)   a.   The woman who [Ø saw the burglar] is here.

        b.   The woman who [the burglar saw Ø] is here.

(34)   a.   The woman who the police reported
             (*that) [Ø saw the burglar] is here.[18]

        b.   The woman who the police reported
             (that) [the burglar saw Ø] is here.

As can be seen in (34a), relativization of the subject of a subordinate clause appears to be unacceptable only if the clause is introduced by an overt complementizer. This suggests that the apparently anomalous correlation between markedness and distribution here may be due to a clash between relativization of subjects and a general parsing strategy similar to the one proposed in Bever and Langendoen (1971), where a non-verbal constituent immediately preceding a verb is analyzed as the subject of that verb. When the complementizer is present in sentences like (34a), such a strategy would yield a semantically incomprehensible analysis. In particular, that would be interpreted as the subject of the verb saw, leaving no constituent in the relative clause that could be interpreted as coreferential with the head (a requirement on all well-formed relative clauses). If such a parsing strategy is indeed involved here, we would expect relativization of the subject of a subordinate to be unacceptable with or without an overt complementizer if the relative clause is immediately preceded by an NP, adverbial or other non-verbal constituent. This is in fact the case, as illustrated by (34c).

(34c)   *The woman who the police $\begin{bmatrix} \text{reported yesterday} \\ \text{told John} \end{bmatrix}$

         (that) [Ø saw the burglar]   is here.

Let us turn now to the second anomalous case, involving the distribution of verb-medial and verb-final clauses in German. Here we

found that the verb medial SVO and OVS orders occur only in main
clauses, while the verb-final SOV order occurs only in subordinate
clauses. Assuming that subordinate clause is a more restricted
context than main clause[19], we have the following situation. In
the case of SVO vs. SOV, the structure with wider distribution, SVO,
is not unmarked relative to the structure with narrower distribution, SOV. In this case, the apparent lack of correlation between
markedness and distribution is no doubt attributable to the fact
that German, due to general processes of language change, developed
different orders in main and subordinate clauses. Any attempt to
explain this development is beyond the scope of this paper[20]. It
should be noted, however, that while the situation here is inconsistent with the first part of the hypothesis in (2), which states
that forms with wider distribution in a language will be typologically unmarked with respect to corresponding forms with narrower distribution, it is not inconsistent with the second part of the hypothesis, which states that typologically unmarked forms will have
wider distribution in a given language than corresponding marked
forms, since the forms in question here, SVO and SOV, are not in a
typological markedness relation at all.

With respect to the other two alternating word orders in German,
SOV vs. OVS, the situation is different. Here the two structures
are in a typological markedness relation and the unmarked form, SOV,
which occurs only in subordinate clauses, has a narrower distribution than the marked form, OVS, which occurs in main clauses. Thus,
the facts here appear to be inconsistent with both parts of the hypothesis in (2). The unmarked form, SOV, does not have wider distribution than the marked form, OVS, and the form with wider distribution,
OVS, is not unmarked relative to the form with narrower distribution,
SOV, with which it alternates. It appears that in this case, however, the apparent lack of correlation can be explained in terms of
the pragmatic function of the forms in question. The lack of correlation between distribution and markedness relations in this case is
a result of the occurrence of the OVS order in main clauses. This
particular order serves the function of placing an object topic or
focus in a prominent, i.e. sentence initial, position, a function
which can not be served by either of the alternative orders, SVO or
SOV. Moreover, if we construe context broadly to embrace the discourse and pragmatic environment of sentences, then it is not true
that OVS, the marked member of the typological markedness relation,
has a wider distribution than SOV, the corresponding unmarked member
of the relation, since OVS occurs in main clauses only when the
object of the sentence is topic or focus, whereas SOV always occurs
in subordinate clauses, regardless of the discourse or pragmatic
context.

Difference in communicative function between the forms in question also appears to be responsible for the apparently anomalous
distribution of dislocational and non-dislocational structures in

English. Dislocational sentences are well suited to the task of encoding topic-comment relations, because they make the distinction between topic and the rest of the sentence structurally explicit. They are, in particular, ideally suited for introducing and reintroducing topics into a discourse. However, new topics are rarely if ever introduced in a subordinate clause. Moreover, in the case of some subordinate clauses, namely relatives, the topic-comment structure of a dislocational sentence actually conflicts with that of the subordinate clause, since the topic of the relative clause is always the relativized NP (cf. Gundel 1974, Kuno 1976,). Thus, the lack of correlation between typological markedness and distribution in this case is due to the specific communicative function of the forms in question, i.e. to the fact that dislocated structures are typically used for introducing or reintroducing a topic in English and this function is incompatible with that of subordinate clauses, from which the typologically unmarked dislocational sentences are excluded.[21]

Thus, in each of the three cases where the typologically unmarked form does not have the wider distribution we have found on closer examination that there are independent factors which override the general communicative advantage of the typologically unmarked forms.

In addition to the observed instances of apparently contrary correlations between markedness and distribution--where the marked rather than the unmarked member of a typological markedness relationseems to have the wider distribution--there are situations, as in the German word order example, where there appears to be no correlation at all. Such lack of correlation between markedness and distribution can arise in two ways: where there is a markedness relation without any associated difference in distribution, or where there is a difference in distribution without any associated markedness relation. Cases appear to exist of both of these two types.

A possible case of the first type might be suggested by the relation between continuous and discontinuous prepositional phrase structures in English questions and relative clauses. These two forms, as exemplified by the alternations in (35), might at first glance appear to be in free variation in all contexts in this language.

(35)   a.   To whom did you give the book?
             Who(m) did you give the book to?

        b.   This is the man to whom you gave the book.
             This is the man who(m) you gave the book to.

There is, of course, a clear typological markedness relation between these two types of prepositional phrases, since there are numerous

languages that have the continuous type but not the discontinuous
one--e.g. Arabic, German, Spanish, Thai--but no language in which
all prepositional phrases are discontinuous. Thus we would appear
to have a case here where the marked and the unmarked members of a
typological markedness relation have exactly the same range of distribution in a language.

If this were the whole story, it would suffice to demonstrate
that the second clause of the general correlation between markedness
and distribution stated in (2) is false--that is, that it is not the
case that if there is a typological markedness relation between two
alternants in a language, the unmarked alternant always has a wider
distribution than the marked. It might still be true, though, that
the marked member never has a wider distribution than the unmarked.

But, in any event, the data indicated in (35) clearly do not
tell the whole story here. First, as can be seen from (36), the
alternation between continuous and discontinuous prepositional
phrases in English is not confined to the context of questions and
relative clauses.

(36)   a.   I gave my book to that guy over there
        b.   To that guy over there I gave my book
        c.   That guy over there I gave my book to

Moreover, as can be seen from (36), there are triplets of synonymous sentences in English such that two of them contain continuous
prepositional phrases and only one contains a discontinuous one.
Other things being equal, therefore, this would suffice to demonstrate that the continuous phrase, the typologically unmarked member
of the pair, does indeed occur in a wider range of contexts than the
discontinuous phrase, the marked member of the relation.

Even clearer evidence of the actual inequality of distributional ranges of these two structures is provided by the kind of
data exemplified in (37).

(37)   a.   Whenever you give a book to that guy over there, he always
             forgets to return it.

        b.   That guy over there, whenever you give a book to him, he
             always forgets to return it.

        c.   *That guy over there, whenever you give a book to, he
             always forgets to return it.

Thus, whereas continuous prepositional phrases are always possible
in all contexts, there are certain contexts where their discontinuous counterparts are absolutely prohibited--in particular, in all
such contexts as in (37c), where anaphoric ellipsis is precluded by

general constraints for English on the linear and structural relationships that must hold between ellipsis sites and their possible
antecedents.

It can be seen on closer examination then that what appeared at
first to be an equality of distributional ranges for the marked and
unmarked members of a typological markedness relation actually constitutes a perfectly regular correspondence of the generally expected type, with the unmarked member having a wider range of distribution than the marked member. More important, though, is the general
methodological moral that this case illustrates.

Thus if we look only at some <u>subset</u> of the instances of an alternation pattern (e.g., only the interrogative and relative instances of prepositional phrase alternations) rather than at the
most general characterization of the alternation pattern as a <u>whole</u>
(e.g., [...NP$_i$...[Adp $\emptyset_i$]...]and [... [Adp NP]...]), then we
may arrive at accidental and misleading conclusions about distribution patterns which fail to accurately reflect the grammatical
structure of the language as a whole. To avoid such artificial
skewing, therefore, it is necessary to adhere strictly to the
universal scientific goal of maximal generalization by seeking to
identify the full range of every alternation relation in its most
general form. The same consideration applies as well, of course,
with respect to the identification of the members of typological
markedness relations. To avoid accidental and misleading conclusions, relations must be investigated between maximally general
classes of structures and not between arbitrarily selected
subclasses.

Another type of potential lack of correlation between markedness
and distribution would arise if there were clearly unequal ranges of
distribution for two alternants in some language but no typological
markedness relation between them. This could be the case either
because each of the forms implies the presence of the other or
because neither form implies the presence of the other.

An apparent instance of the first sort involves the alternation
between present and past tense clauses in English. Although these
two structures are in contrast, clearly, in a very wide range of
different contexts, there is at least one context where this contrast is neutralized--optionally for most speakers but perhaps ob-
ligatorily for at least some--in favor of the past tense form. This
is the context of polite interrogation, as in (38b), where a past
tense form is ambiguously interpretable as either present or past,
in contrast to ordinary questions, as in (38a), and most other
situations, where past and present forms always have distinctively
different interpretations.

(38)   a.   <u>Ordinary Questions</u>
              Do you want that report (now/*yesterday)?
              Did you want that report (*now/yesterday)?

        b.   <u>Polite Questions</u>
              Do you want that report (now/*yesterday)?
              Did you want that report (now/yesterday)?

The alternation that is involved here is between the past and non-past forms of clauses. Typologically, however, it seems evident that if a language has a distinctive past form it must also have a distinctive non-past form, and if it has a non-past form it must also have a past form. Thus there could be no typological markedness relation between past and non-past clauses. We would seem to have here a neutralization relation in a particular language between forms that are not in any markedness relation to each other at all.

It should be noted, though, that there are factors which render this case particularly unclear--with respect to both distribution and markedness. Thus the operant contextual distinction--between polite and non-polite discourse--is, in contrast to all of the other contextual distinctions we have referred to, based on factors other than form or meaning. Moreover, with respect to the determination of typological relations, the entities past and non-past, unlike the various other entities that have been investigated, appear to be <u>logically</u> dependent upon each other, and thus could not be related in any possible <u>empirical</u> way. It might be that it is only under such special circumstances as these, therefore, that this type of lack of correlation between markedness and distribution can be observed.

A lack of correlation could also arise, though, if neither of two unequally distributed forms is typologically dependent on the other. There is at least one evidently quite clear case of this sort in phonology. This involves the relationship between the lateral and non-lateral liquids, <u>l</u> and <u>r</u>. There is no typological markedness relation between these two sounds, since there are some languages that have <u>l</u>-sounds but no <u>r</u>-sounds (e.g., Cantonese, Hawaiian, Kiowa, and Klamath) and other languages that have <u>r</u>-sounds but no <u>l</u>-sounds (e.g., Tahitian, Winnebago, Karok, and Santa Clara Tewa). But in some dialects of Latin American Spanish these two sound types stand in a neutralization relation to each other. In fact, moreover, in some parts of Puerto Rico, as shown in (39), the contrast between <u>l</u> and <u>ſ</u> is neutralized in favor of <u>l</u>, and in other varieties, as shown in (40), the contrast is neutralized in favor of <u>ſ</u>. (Canfield, 1981:76).

(39)   <u>singular</u>   <u>plural</u>
        [mal]   [males]   'evil'
        [mal]   [maſes]   'sea'

(40)   <u>singular</u>   <u>plural</u>
          [maʎ]   [males]   'evil'
          [ma ɾ]   [maɾes]   'sea'

The Spanish data here suffice to demonstrate that the general
correlation between markedness and distribution in the form stated
in (2) is false. In other words, regardless of the existence or
nature of any markedness relationship between <u>l</u>-sounds and <u>r</u>-sounds,
the fact that one dialect neutralizes to <u>l</u> and the other to <u>r</u> shows
that it cannot be the case that whenever there is a suspension of
alternation the form that occurs in the position of suspension is
the unmarked member of a typological markedness relation.

But it will be be observed now that since there is in fact <u>no</u>
typological markedness relation between <u>l</u> and <u>r</u> , it is true for
both of the Spanish dialects that the form which occurs in the position of suspended alternation is <u>not</u> a <u>marked</u> form. The Spanish
data would thus be consistent with the revised correlation principle
stated in (41).

(41)   Typological markedness and difference of distribution are
          correlated such that, in a given language with two alterna-
          ting forms A and B, if A has a wider distribution than B,
          then A is not typologically marked relative to B, and if A
          if typologically unmarked relative to B, then A has a wider
          distribution than B.

This principle is consistent not only with the Spanish data but
also with all of the other examples of various distributional relations that we have examined here. Though it specifies a weaker correlation than (2) does between markedness and distribution, it nevertheless generates a still quite restictive empirical claim about the
range of possible variability of human languages. Thus, for
example, it follows from (41) that there could be possible languages
exhibiting the types of distribution patterns represented in (42a-c)
but no possible language with the pattern represented in (42d).
(Subscripted <u>u</u> and <u>m</u> signify the unmarked and marked members of a
typological markedness relation, respectively, and subscripted <u>o</u>'s
signify alternants that are not in any markedness relation to each
other.)

(42)   a. A_u/X A_u/Y b. A_o/X A_o/X
            B_m/X $*B_m/Y$ B_o/X $*B_o/Y$

        c. A_o/X $*A_o/Y$ *d. A_u/X $*A_u/Y$
            B_o/X B_o/Y B_m/X B_m/Y

This principle of correlation in (41) seems in fact to have

greater intuitive plausibility than the more restrictive correlation principle stated in (2), which would predict the non-existence of patterns (42b) and (42c) as well as (42d), and would thus be falsified by the cited facts about l-r alternations in Spanish. For if the existence of a typological markedness relation between two structures is due to a difference in their relative effectiveness for purposes of human communication in conjunction with the evolutionary favoring of more effective devices over less effective ones, then when there is no markedness relation at all between two structures, this should mean that there is no significant difference between them in their communicative value or utility and no reason, therefore, for any language to favor the use of one over the other.

In such an event, in other words, there would be no gain or loss if an alternation is suspended or not, and, if suspended, whether one form or the other occurs in the position of suspension. In such cases, then, since there is no reason for languages to do one thing rather than another, different languages should be **expected** to do **different** things—which is exactly what is found with respect to the l-r situation in Spanish and other languages. The correlation principle in (41), then, would appear to appropriately reflect the fundamental distinction in natural language between those aspects of linguistic structure that are functionally determined and those that are underdetermined by functional factors and hence free to vary arbitrarily from one language to another.

NOTES

[1]We use the terms alternant and alternation here in a more general sense than usual to include forms that are in free variation or complementary distribution as well as those that are in contrast with each other. Thus, we speak of suspended alternation in cases of absence of free variation as well as absence of contrast.

[2]The relation "is interpreted as" is a non-symmetric relation indicating the value of a linguistic form, in the sense of Saussure (1915[1959]). We take the assertion that A<(A,B) to be equivalent to the assertion that A=B.

It should be noted that when we add interpretation to schema (3c) we get two distinct relations:

(i) Simple defective distribution   (ii) Neutralization

      A/X   A/Y   A/X   A/Y
      B/X   *B/Y   B/X   *B/Y
      A<A   A<A   A<A   A< [A,B]
      B<B   B<B

[3]To allow us to investigate each of the logical possibilities,
we restrict our attention here to relations involving only two contexts. Extension to more than two contexts is discussed in section
3.6.

[4]Here, as elsewhere in this paper, we are assuming for purposes
of discussion, equal ranges of distribution in contexts other than
those explicitly discussed.

[5]Some linguists (e.g. Trubetzkoy 1939[1969:79-80]) have analyzed the stop that occurs after /s/ in English as neither the
voiced nor the voiceless phoneme, but rather as an archiphoneme that
has only the features common to both.

[6]There might be a few Australian languages that constitute
exceptions to this generalization. In Ruhlen (1975), based on a
sample of 700 languages, there are only four languages whose cited
phonemic inventories lack voiceless obstruents: Dyirbal (NE
Australia, Pama-Nyungan), Gudandji (Kutandji) (NC Australia, Pama-
Nyungan), Ngarndji (NC Australia, Pama-Nyungan), and Yanyula (NC
Australia, Pama-Nyungan). In the UCLA Phonological Segment Inventory
Database (UPSID), a sample of 317 languages, there are only two
languages cited without voiceless obstruent phonemes: Alawa (NC
Australia) and Bandjalang (E Australia).It seems to be the case,
however, that at least some of these six languages (and possibly all
of them) have voiceless obstruents phonetically, and that the choice
of voiced obstruent symbols for their obstruent phonemes may be in
at least some cases merely arbitrary or a matter of extra-linguistic
convention among Australianists. Thus, concerning Yanyula, Kirton
(1967:18) says: "The series of stops vary according to voicing,
voicelessness and voiceless aspiration, this latter feature being
more apparent in speakers who have had most contact with English.
These allophones occur in free variation but the tendency has been
noted for voicing to become frequent between vowels and at the onset
of a stressed syllable in word medial position." Similarly, with
respect to Bandjalang, Cunningham (1969) gives [p] and [pʰ] along
with [b] as allophones of the labial stop phoneme represented as /b/,
and [k] along with [g] as an allophone of the velar stop phoneme /g/.
In the case of Alawa, whose phonemic inventory is shown in UPSID
without any voiceless obstruent symbols, Ruhlen (1975:155), citing
the same source as UPSID (Sharpe 1972), shows a whole series of
voiceless stop phonemes (p, t, ṭ, c, k) in contrast with a parallel
series of voiced phonemes (ᵐb, ḍ, d, ⁿj, g). With respect
to Gudandji, Aguas (1968) describes each of the obstruent phonemes
of the language (b, d, ḍ, dj, g) as a "lightly voiced stop, unaspirated in all instances." It is possible, though, that there
would not be any perceptible difference between a "lightly voiced"
unaspirated stop and voiceless unaspirated stop. Concerning Dyirbal
and Ngarndji, we have thus far been unable to find any primary information about their obstruent sounds. (Dixon's (1972) lengthy

grammar of Dyirbal contains no statement at all about the voicing characteristics of obstruents.) The strongest candidate for excep-tionality to the generalization that we have been able to find is Yidiny, another Australian language that is not included in either the UPSID or Stanford samples. Thus, in his grammar of this lan-guage, Dixon (1977:32) says: "Stops [the only obstruents] are almost always voiced. Partly voiced allophones are sometimes encountered word-initially (most commonly, the word begins an intonation group); words cannot end in a stop. It is, in fact, normal for the glottis to be vibrating throughout the articulation of a Yidiny word; thus, in one sense, 'voiced' is (for this language) the unmarked value of the phonetic opposition 'voiced/voiceless'."

[7]Actually there is a third context where only unreleased stops occur, namely, word-medial, syllable-final position, as in <u>actor</u>. Therefore, this is really a multiple context situation of the sort discussed in 3.6.

[8]There is, of course, also a situation where the alternation is suspended in favor of the <u>V NP   Particle</u> structure, namely, when the NP is an unstressed pronoun. However, although not all NPs can occur in the <u>V Particle NP</u> structure in English, it is still true that this structure can occur in any context in which <u>V NP Particle</u> occurs.

[9]It will be observed that if the terms for comparison here were taken to be clauses with initial question phrases and clauses with non-initial question phrases, then there would be no markedness relation between the two structures at all -- since there are evi-dently no languages in which all question phrases are always clause-initial and also no languages in which question phrases are never clause-initial. Under such an analysis, the general correlation in (2) would be violated, but it would still not be the case that a structure that is marked has a wider distribution than a correspon-ding unmarked structure. Thus the revised correlation principle to be given in (41) below would still hold.

[10]Thus, for example, with a pronominal possessor, like <u>wo</u> 'I' in <u>wo de mugin</u> 'my mother', the marker <u>de</u> can be optionally omitted as in the synonymous phrase <u>wo mugin</u>. In most other types of geni-tival phrases the <u>de</u> is obligatory (see Chao 1968:289).

[11]A hypothetical example of neutralization by compromise to an unamrked rather than a marked form would be provided by a language that had <u>s</u> and <u>z</u> in contrast initially and neutralization of <u>s</u> and <u>z</u> to <u>t</u> finally, where <u>t</u>, which is typologically unmarked relative to <u>s</u> and <u>z</u>, does not occur initially. It appears to be the case so far that there are no real situations of this type.

[12]It should be noted that this generalization holds only for

adverbial and coordinate clauses. In Chinese, as presumably in all languages, a single NP can occur as an answer to a question.

[13]This assumption is based on the fact that every sentence has a non-subordinate clause, but not every sentence has a subordinate clause.

[14]Thus, it is normally the case even in the most richly inflected and cross-referential languages that verbs used as nominals (gerunds) or as adjectival modifiers (participles) exhibit few if any of the distinctions in argument and adverbial selection that they manifest when functioning as predicates. Such absence of verbal distinctions in person, number, gender, tense, aspect, etc., is evidently complete, moreover, in at least some contexts in all languages--for example, in compounds of the sort illustrated below.

Let's have a looksee. (*lookedsee, *looksaw, *looksees)
I planted forget-me-nots. (*forgot-me-nots, *forgets-me-nots,
                            *forgetting-me-nots)

[15]For example, after syllable-initial stops only the tap can occur, as in t[ɾ]es 'three'.

[16]It also should be noted that there is an additional context--stem-initially before a consonant--where both /s/ and /z/ are defectively distributed. Thus, to characterize all the distributional facts about these two sounds, at least four contexts must be considered.

[17]It might be of interest to note that Trubetzkoy (1939[1969: 82-83]) takes these facts as evidence that /z/ is the unmarked member of the /s/-/z/ opposition in German, since it represents the archiphoneme in initial position, where more contrasts are maintained than in final position. This is a clear indication of Trubetzkoy's view that a) markedness is language-specific and b) neutralization (including defective distribution) is the basis for determining which member of an opposition is marked.

[18]It should be noted that some people find (34a) with the complementizer to be not clearly ungrammatical. A much clearer case is provided by sentences with other complementizers, such as *The woman who the police wondered whether saw the burglar is here.

[19]See footnote 13.

[20]For general hypotheses concerning the nature and causes of such developments in German and other languages, see Vennemann (1974) and his other studies cited therein.

[21]Similar points have been made by Vennemann in a number of his general studies of word order change. For example, in Vennemann

(1974):362) he notes that languages which have changed their main clause order from XV to Topic VX "may retain the XV pattern in subordinate clauses, where topicalization plays a smaller role than in main clauses."

REFERENCES

Aguas, E.F. 1968. Gudandji. Pacific Linguistics, A, 14, 1-20.

Bever, T.G. and D.T. Langendoen. 1971. A dynamic model of the evolution of language. Linguistic Inquiry. 2.4:433-464.

Canfield, D.L. 1981. Spanish pronunciation in the Americas. Chicago: The University of Chicago Press.

Chao, Yuen Ren. 1968. A Grammar of Spoken Chinese. Berkeley: University of California Press.

Comrie, Bernard. 1984. Why linguists need language learners. In Language Universals and Second Language Acquisition, ed. by W.E. Rutherford. Amsterdam: John Benjamins Publishing Co.

Cunningham, M.C. 1969. A description of the Yugumbir dialect of Bandjalang. (University of Queensland Faculty of Arts Papers, 1/8.) Brisbane: University of Queensland Press.

De Burca, Seán. 1970. The Irish of Tourmakeady. Dublin: Dublin Institute for Advanced Studies.

Derbyshire, Desmond C. 1977. Word order universals and the existence of OVS Languages. Linguistic Inquiry 8.3:590-99.

Dixon, R.M.W. 1972. The Dyirbal language of North Queensland. Cambridge: Cambridge University Press.

----------. 1977. A Grammar of Yidin. London: Cambridge University Press.

Eckman, Fred. 1977. Markedness and the contrastive analysis hypothesis. Language Learning 27.2:315-30.

Ferguson, Charles A. 1961. Assumptions about nasals: a sample study of phonological universals. In Universals of Language, ed. by J.H. Greenberg, 2nd edition, 1966. Cambridge, Mass: MIT Press.

Greenberg, Joseph H. 1966. Language Universals. The Hague: Mouton.

----------. 1965. Some generalizations concerning initial and final consonant clusters. Linguistics 18:5-32.

Gundel, Jeanette K. 1974. The role of topic and comment in linguistic theory. Ph.D. Dissertation. University of Texas at Austin. Reproduced by Indiana University Linguistics Club, 1977.

Harris, James W. 1969. Spanish phonology. Cambridge, Mass: MIT Press.

Houlihan, Kathleen and Gregory K. Iverson. 1979. Functionally constrained phonology. In Current Approaches to Phonological Theory, ed. by D.A. Dinnsen, pp. 50-73. Bloomington: Indiana University Press.

----------. 1980. On determining the markedness of phonological
    segments, Minnesota Papers in Linguistics and Philosophy of
    Language, ed. by N. Stenson, pp. 73-83.
Jakobson, Roman. 1968. Child language, aphasia and phonological
    universals. The Hague: Mouton.
Keenan, Edward L. and Bernard Comrie. 1977. Noun phrase accessi-
    bility and universal grammar. Linguistic Inquiry 8.1:63-100.
Kim-Renaud, Young-Kee. 1974. Korean Consonantal Phonology. Ph.D.
    dissertation. University of Hawaii at Manoa.
Kirton, Jean F. 1967. Anyula phonology. Pacific Linguistics, A,
    10, 15-28.
Kuno, Susumu. 1976. Theme, rheme and the speakers' empathy. In
    Subject and Topic, ed. by Charles N. Li, New York: Academic
    Press, pp. 417-444.
Ladefoged, Peter. 1982. A Course in Phonetics, 2nd edition.
    New York: Harcourt Brace Jovanovich.
Lehtinen, Meri. 1963. Basic Course in Finnish. Bloomington:
    Indiana University Publications.
Mattoso Camara, Jr. J. 1972. The Portuguese Language.
    Translated by A.J. Naro. Chicago: The University of Chicago
    Press.
Moulton, William G. 1962. The Sounds of English and German.
    Chicago: The University of Chicago Press.
Nunez, Rafael A. 1977. Fonologia del espanol de Santo Domingo.
    Ph.D. Dissertation. University of Minnesota.
Quilis, Antonio and Joseph A. Fernandez. 1966. Curso de fonetica
    y fonologia espanolas. Madrid: Consejo Superior de Investi-
    gaciones Cientificas.
Ruhlen, Merrit. 1975. A Guide to the Languages of the World.
    Stanford: Stanford University Language Universals Project.
de Saussure, Ferdinand. 1915. Course in General Linguistics.
    English translation. New York: Philosophical Library, 1959.
Trubetzkoy, N.S. 1939. Principles of Phonology. English trans-
    lation by C.A.M. Baltaxe. Berkeley: University of California
    Press, 1969.
UPSID: UCLA Phonological Segment Inventory Database. 1981. UCLA
    Working Papers in Phonetics 53.
Vennemann, Theo. 1974. Topics, subjects and word order: from
    SXV via TVX. In J.M. Anderson and C. Jones, eds., Historical
    Linguistics, vol. 1, pp. 339-376. Amsterdam: North Holland
    Publishing Co.
Whitney, William Dwight. 1889. A Sanskrit Grammar.
    Leipzig: Breitkopf and Härtel.

MARKEDNESS AND THE BINDABILITY OF SUBJECT OF NP

Wayne Harbert

Cornell University

There is a set of domains, D_1, within which pronouns are required to be free--that is, within which they may not be coindexed
with any c-commanding argument NP. Coreference is possible between
a pronoun and an NP which occurs outside of its D_1. This is
illustrated by the contrast between (1a) and (1b), where I adopt the
convention of underscores to indicate intended coreference.

1a. *[D_1 John hurt him]

 b.   John expected [D_1 Mary to hurt him]

There is a set of domains, D_2, within which anaphors such as reciprocals and reflexives must be bound--that is, coindexed with a c-
commanding argument NP. The antecedent of an anaphor may not occur
outside of the domain D_2 containing that anaphor. This is illustrated by the contrast between (2a) and 2b).

2a. [D_2 John hurt himself]

 b. *John expected [D_2 Mary to hurt himself]

Much attention has been devoted in recent Extended Standard Theory
literature to formulating general definitions of the domains in
question. The formulations advanced have in general incorporated
the assumption that D_1, the binding domain for pronouns, and D_2,
the binding domain for anaphors, are identical. This is not an
unreasonable assumption, since pronouns in fact usually are subject
to an interpretation of obligatory disjoint reference with respect
to a c-commanding NP if an anaphor occurring in the same position
could be bound to that NP. Consider, for example, the complementarity of asterisks in (1) and (2).

139

Such formulations, however, have invariably left unaccounted for
a number of residual cases in which the usual complementarity between anaphors and bound (proximate) pronouns is not found. Moreover, in at least a subset of the constructions in question, the
noncomplementarity appears to persist across languages--a fact which
has led Huang (1982) to claim that the definitions of binding domains for pronouns and anaphors in universal grammar, while over-
lapping substantially, do in fact differ with respect to one crucial
provision.

In the following, I will review some of the evidence from English
and other languages which demonstrates the superior factual coverage
of Huang's two-part characterization of binding domains over the
single-domain formulations advanced in earlier work. I will also
point out a fact about the distribution of anaphors in the position
of subject of NP in English which appears to resist synchronic
explanation even under the assumption of distinct binding domains.
I will show, however, that this fact can be explained, assuming
Huang's formulation of binding domains, as the natural result of a
linguistic change occasioned by pressure to eliminate markedness.
The same explanation will be shown to extend to some independent but
strikingly similar developments in other IndoEuropean languages.

Consider first the two competing definitions of binding domains
presented in Chomsky (1981), which I have paraphrased as (3a) and
(3b). We may take these as representative of the uniform-domain
hypotheses.

3.   Principle A: An anaphor must be bound in its binding
                     category
     Principle B   A pronoun must be free in its binding
                     category

     a.   The Binding Category (D) for α is the minimal NP or
          S containing α and a governor of α .[1]

     b.   The Binding Category (D) for α is the minimal
          category containing α and a SUBJECT accessible
          to α .[2]

Taken in conjunction with principles A and B of the binding
theory as stated in (3), definitions (3a) and (3b) both yield iden-
tical--and correct--predictions with respect to the examples in (1)
and (2). In (1b), for instance, him is governed by hurt. Its
binding category under definition (3a) is therefore the bracketted
complement D_1, which is the lowest S containing the pronoun and
its governor. D_1 is also the binding category for him under definition (3b), since it is the lowest category containing that pronoun
and a subject, Mary, which c-commands it. Thus, both definitions

correctly predict that <u>him</u> may corefer with a c-commanding NP outside
D_1. More generally, both hypotheses make identical (and correct)
predictions about the distribution of overt anaphors and pronominals
for virtually all NP positions within clauses.[3] For a complete
demonstration of this, the reader is referred to Chomsky (1981).

For NP positions within noun phrases, on the other hand, the predictions of the two diverge sharply. For the sake of brevity, I will
restrict my attention to the position of subject of NP--i.e., the
position occupied by NP_3 in (4).

4.   [$_{NP1}$ He] believed [$_{NP2}$ [$_{NP3}$ the man's] [$_{\bar{N}}$ stories]]

NP_3 is governed within NP_2 by <u>stories</u>, the head of NP_2, since
no maximal phrase boundary intervenes between them. Thus, under
definition (3a), NP_2 is the binding category for NP_3. It is
consequently predicted that an anaphor occurring in the position of
NP_3 must find an antecedent within NP_2--and therefore, since
NP_2 contains no appropriate antecedent position, that NP_3 cannot
be realized as an anaphor at all. (3a) predicts further that, if it
is a pronoun, NP_3 may corefer with NP_1; it will still be free
within its binding category, NP_2, as required.

Under definition (3b), on the other hand, an NP constitutes a
binding category for a position contained within it only if the NP
also contains a SUBJECT of, and is therefore not c-commanded by the
SUBJECT of, NP_2.[4] Thus, under (3b), S, not NP_2, is the binding
category for NP_3, and it is predicted that NP_3 may be an anaphor
bound to NP_1, and conversely that if NP_3 is a pronoun it is
necessarily disjoint in reference with NP_1.[3]

We see, therefore, that (3a) and (3b) make opposite predictions
about the possible realization and interpretation of the position of
NP_3 in (4). When we turn to the facts of English to determine
which of these predictions is borne out, we get conflicting results.
Consider (5a) and (5b).

5a.   [$_S$ <u>They</u> sold [$_{NP}$ <u>each other's</u> pictures]

 b.   [$_S$ <u>They</u> sold [$_{NP}$ <u>their</u> pictures]

As these examples show, the usual parallelism between anaphor binding
and obligatory disjoint reference interpretation of pronouns breaks
down precisely where the predictions of (3a) and (3b) diverge.
Neither of the uniform-domain definitions in (3) can account for
both of the examples in (5). (3a) characterizes NP as the binding
category for the underscored position in (5a, 5b). Thus, it correctly predicts the opacity of NP to obligatory disjoint reference
in (5b), but it incorrectly predicts that NP is also opaque to

anaphor binding, contrary to the facts of (5a). (3b), on the other
hand, characterizes S as the binding category for the underscored
positions in (5a, 5b), since they are not c-commanded by the subject
of NP. Therefore it correctly predicts the transparency of NP in
(5a) to anaphor binding, but it incorrectly predicts that there
should be obligatory disjoint reference between they and their in
(5b).

It is of course possible to devise auxiliary assumptions which
will reconcile the facts of (5) with either of these definitions of
binding categories. Taking (3a) to be correct, for example, we
might say that (5a) reflects a marked relaxation of binding principles in English in the case of anaphors in NP subject position.[5]
Alternatively, taking (3b) to be correct, we might assume that under
certain circumstances a special provision of English grammar defines
a smaller binding domain for pronouns than that defined by the binding principles of universal grammar, thus accounting for (5b).[6]

Huang observes, however, that such solutions fail empirically,
since they predict incorrectly that the noncomplementarity of prox-
imate pronouns and anaphors reflected in (5) is a marked property of
English, when in fact it obtains in other languages as well.
Compare Huang's Chinese examples in (6) with the corresponding
English examples in (5).

6a. Zhangshan kanjian-le[$_{NP}$ ziji-de shu]
     Zhangshan$_i$ see ASP   self$_i$ POSS book
     'Zhangshan saw his (own) book.'

 b. Zhangshan kanjian-le [$_{NP}$ ta-de shu]
     Zhangshan see   ASP   he$_i$ POSS book
     'Zhangshan saw his (own) book.'

In view of this, and of certain conceptual problems with (3a) and
(3b), Huang proposes that these definitions should be abandoned in
favor of a formulation which characterizes binding domains for
pronouns and binding domains for anaphors differently. The particular formulation he adopts is given in (7).

7. β is a binding category for α iff:

    β is the minimal category containing α , a governor
    of α , and a SUBJECT which, if α is an anaphor, is
    accessible to α .

Huang justifies this formulation in detail. For present purposes,
we may concentrate on its central feature: (7) required only binding categories for anaphors, not those for pronouns, to contain
accessible SUBJECTS.

(7) duplicates the predictions of (3a) and (3b) for all cases
where those two converge. However, it also accounts for the asymmetry in (5) and (6), which does not follow from either (3a) or (3b).
Under definition (7), NP is the binding category for the pronoun in
(5b), since NP contains a SUBJECT--the pronoun itself, if we take
SUBJECT to be defined as in note (2). Thus, the pronoun need be
free only within NP. However, under (7), NP is not a binding
category for the anaphor each other's in (5a), since the SUBJECT of
NP does not c-command the anaphor, hence is not accessible to it in
the sense of note (2).[7] The anaphor may consequently be bound
outside of NP. Therefore, (7) predicts the noncomplementarity
between proximate pronouns and anaphors in (5) and (6).

Huang's two-part definition of binding categories is further
supported by similar instances of noncomplementarity in other languages. In New Testament Greek, for example, we find a distribution essentially identical to that in (5) and (6); as in English
and Chinese, both pronouns and anaphors in the position of subject
of NP in Greek can corefer with the subject of the clause containing that NP. This is illustrated in (8a) and (8b).

8a. Kaì (autós) elegen autoĩs en [$_{NP}$ tẽi didakhẽi autoũ]
     and he$_i$ spoke-to-them in the teaching his$_i$
     'And he spoke to them in his teaching.' (Mk 4:2)

b.   áphes toùs nekroùs thapsai [$_{NP}$ toùs heautõn nekroùs]
     let   the dead   bury   the self's$_i$ dead
     'Let the dead bury their dead.' (Mat. 8:22)

I expect that similar examples will not be hard to find elsewhere.

In Harbert (1983) I argue that (7) can be extended to cover more
complex instances of asymmetry between pronouns and anaphors in
other languages, as well. Thus, in terms of coverage of the facts
it appears to represent an improvement over earlier formulations of
binding domains. However, there are some phenomena involving the
behavior of the position of subject of NP with respect to binding
which remain unpredicted even under the assumption of (7). First,
while reciprocals may appear in that position in English, bound to a
higher subject, as in (5a), reflexives do not. Consider the ungrammatical sentence in (9).

9.   *She sold [$_{NP}$ herself's pictures]

Now, the ungrammaticality of (9) does not appear to be a fact which
we would wish to account for by further qualification of the operation of binding conditions in English. For example, we would not
want to stipulate that the binding domains of reflexives and reciprocals are differently defined, with NP constituting a binding category

for the former, but not the latter. Besides being ad hoc, such an
account would fail empirically.[8] (10) shows that reflexives inside
of NPs in English can in fact find their antecedents outside of those
NPs.

  10. <u>They</u> sold [$_{NP}$ pictures of <u>themselves</u>]

A more plausible approach to (9) therefore would be to claim that it
is not ruled out by binding theory at all, but is bad simply because
English has no lexical items like <u>herself's</u>—that is, that the
reflexive paradigms are defective, lacking genitive forms.

    This does not mean, however, that binding theory has no respon-
sibility for accounting for (9). We should hope that an optimally
formulated binding theory will provide us with a way of explaining
this gap in the paradigm. There are, it seems, not just one but two
such facts in need of explanation. As illustrated by (11), English
did once have a strictly reflexive possessive form, <u>sīn</u>, which has
been lost. Its loss was independent of the disappearance of the
other Germanic reflexive forms in pre-English, which took place
several hundred years earlier.

  11. and him <u>Hrōðgar</u> gewat in [$_{NP}$ hofe <u>sīnum</u>]
        and him(self) Hrothgar$_i$ betook to house self's$_i$
        'And Hrothgar betook himself to his house.' (Beow. 1236)

Thus, we must ultimately account for both the loss of the OE <u>sīn</u> and
the failure to create a replacement for it when English subsequently
introduced the <u>self</u>-reflexive paradigm.

    A second and related problem remaining for (7) is this; there
are languages in which, unlike English, Chinese and Greek, and
contrary to the predictions of (7), pronouns occurring as subject
of an NP <u>must</u> be understood as disjoint in reference with the
subject of the clause containing that NP. Icelandic is such a
language. In (12), where coreference is intended with the higher
subject the reflexive must be used. Compare (12) with (5b).

  12. [$_S$Jón rétti Haraldi [$_{NP}$ <u>sín</u>/ *hans föt]
        John$_i$ handed Harold   self's$_i$ *his; clothes
        'John handed Harold his clothes.' (Thrainsson 1976a)

The same appears to have been true of Gothic and Latin. Compare
(13a) and (13b) with (5b).[9]

  13a. jah (<u>is</u>) qaþ im [$_{NP}$ in laiseinai <u>seinai</u>/ (*<u>is</u>)]
         and he$_i$ spoke to-them in teaching self's$_i$ *his$_i$
                              (GK has pron.)
         'And he spoke to them in his doctrine.' (MK 4:2)

b. (is) dixit [$_{NP}$ discipulis suis/ (*eius)]...
   he$_i$ said   to disciples self's$_i$ *his$_i$ (GK has 0)
   'He said to his disciples...'   (Lat Vulg. Mat 26:36)

Thus, it appears that in some languages NP functions as the binding category for a pronoun occurring in its subject position, while in others it does not. Icelandic, Gothic and Latin are of the latter type. Greek, Chinese and English are of the former type. To this group we may also add Japanese, as illustrated by (14).

14. John-wa [$_{NP}$ kare-no hon] -omot-te ki-ta
     John$_i$ TOP   he -POSS book-ACC carry-GER-come PAST
     'John brought his book'   (Oshima, 1979)

There are two possible ways of approaching such a difference between languages. First, we could assume that the definition of binding categories is parametrized in such a way that individual languages are free to decide whether or not NPs constitute binding categories for pronouns. Accordingly, English, Chinese, Greek, and Japanese on the one hand, and Icelandic, Gothic, and Latin, on the other hand, would simply reflect the choice of distinct but equally valued options.[10]

Alternatively, we could suppose that the two patterns are not equally valued, but that one of them exists in marked contravention of the binding conditions of universal grammar. The question is whether there is evidence on the basis of which we can decide between these possibilities.

By way of answering this question, consider another difference between English and Icelandic, represented in the sentences in (15).

15a. He believes [$_S$ me to have seen him/ *himself]

   b. Hann telur [$_S$ mig hafa séð *hann/sig]
       He$_i$ believes me to-have seen *him$_i$ self$_i$ (Andrews,
1976)

These two examples show that the two languages do not differ solely with respect to the interpretation of pronouns in the position of subject of NP. In Icelandic, unlike English, a pronoun in nonsubject position in an infinitive complement must be understood as disjoint in reference not only with the subject of its own clause but with the subject of the higher clause as well. Again, where coreference with the higher subject is intended the reflexive must be used. Moreover, it is apparently not accidental that the two languages differ from each other in both respects. In New Testament Greek and Japanese, pronouns in analogous positions in infinitive complements do admit coreference with the higher subject, as illustrated by the examples in (16). That is, these languages pattern once again with English.[11]

16a. kaì ērōtēsen autòn_j hapan to plēthos... [_S PRO_j apeltheîn
      and asked   him all the population_j   to-go
      ap' autōn (*heautōn)]
      from them_i *selves_i
      'And all the population...asked him to go from them.'
                                                   (Luk 8:37)

  b. John-wa Mary_j -ni [_S PRO_j kare-ni denwa-o kake]-sase-ta
     John_i -TOP Mary-DAT him_i -DAT phone-ACC make cause past
     'John made Mary make a phone call to him.' (Oshima, 1979)

In Latin and Gothic, on the other hand, pronouns in corresponding positions apparently did not admit coreference with the higher
subject; when coreference is intended the reflexive appears, as in the
examples in (17).[12] That is, these languages again pattern with
Icelandic.

17a. qui noluerunt [_Sme regnare super se / (*eos)]
     who not-wanted   me to-rule over selves_i *them_i
     'who did not want me to rule over them' (Vulg. Lk 19:27)

  b. ai -ei ni wildedun [_S mik_j iudanon ufar sis / (*im)]
     who_i compl. not wanted me to-rule over selves_i *them
     'who did not want me to rule over them' (GK has Pron.)

In view of the fact that languages appear to fall into identical groups
with respect to both constructions, it is clear that both of the
observed differences between English and Icelandic are to be related to
a common cause. Note further that none of the versions
of binding theory considered so far predicts the Icelandic pattern in
(15b). (3a), (3b) and (7) all converge in defining the complement
clause as the binding category for the pronoun in (15b), and therefore predict incorrectly that it should be required to be free only
within that clause. No matter which of these definitions we attri-
bute to universal grammar, therefore, we are led to conclude that (15b)
reflects marked nonobservance of some provision of UG.

There are a priori a number of possibilities for characterizing the
observed differences between English and Icelandic. We could suppose, for example, that Icelandic, unlike English, observes trun-
cated versions of principles A and B, which do not make reference
to binding categories.

18a. Principle A': An anaphor must be bound

  b. Principle B': A pronoun must be free

According to this formulation, a pronoun would be subject to an interpretation of obligatory disjoint reference in Icelandic with re-

spect to any c-commanding NP, whether inside or outside of its
binding category. This would capture the disjoint reference facts
of (12) and (15b). Such a characterization fails, however, when we
look at other facts. As (19) shows, finite indicative clauses remain
opaque to disjoint reference and reflexive binding in Icelandic.[13]

19. J$\acute{\text{o}}$n   veit   a$\eth$ [Mar$\acute{\text{i}}$a elskar hann   /*sig]
    John$_i$ knows that Mary   loves   him$_i$ /*self$_i$
    (Thra$\acute{\text{i}}$nsson 1976)

Thus, binding in Icelandic is in fact limited to certain domains
although these domains are a subset of those observed by binding in
such languages as English. An alternative characterization is required, therefore, which blocks obligatory disjoint reference in
(19) while admitting it in (12) and (15b). In developing such an
alternative, we can exploit the fact that the notion SUBJECT, as
defined in note (2), has two subparts which are hierarchically
ordered relative to each other. Where present, the AGR(EEMENT)
element of finite clauses serves as SUBJECT. Syntactic subject NPs
constitute SUBJECTS only in the absence of AGR. Now, suppose that
the definition of SUBJECT is parametized in such a way that languages may draw their threshholds for SUBJECT-hood at different
points on this hierarchy.[14] In English, both AGR and syntactic
subjects are potential SUBJECTS. Let us assume that this is the
unmarked case. In the (presumably marked) case represented by
Icelandic, on the other hand, only AGR would constitute a potential
SUBJECT. Syntactic subject NPs would not. This would account for
the asymmetry between (15b) and (19). In the latter, the pronoun is
in the domain of a SUBJECT--the AGR element--within the complement.
In the former, it is not, since syntactic subject NPs do not constitute SUBJECTS in Icelandic.

Observe that we need say nothing further about the difference
between English (5b) and Icelandic (12). The pronoun in (5b) is
allowed to corefer with the subject of S because NP has a syntactic subject, hence a SUBJECT, and is therefore a binding category
for the pronoun, as defined in (7). The pronoun in (12) is not
allowed to corefer with the subject of S because NP is not a binding
category for that pronoun. NP again has a syntactic subject NP, but
syntactic subjects do not count as SUBJECTS in Icelandic. Thus, the
hypothesis needed to account for the contrast between (15b) and (19)
yields (12) as well. This explains our observation that languages
differing from English with respect to one of these constructions
seem in general also to differ from English with respect to the
other.

Thus, by hypothesis, both the pattern in (12) and the pattern in
(15b) reflect the results of a single, marked choice from among the
options made available in universal grammar. If this is correct, we
should expect to find evidence for the marked status of these con-

structions, in diachrony, for example; they should exhibit a tendency to change in the course of time in the direction of the
unmarked patterns claimed to be exemplified by English. Moreover,
since both of the Icelandic-type constructions are held to result
from the selection of the same marked value for SUBJECT, we predict
that if one of the constructions does change in a given language,
the other will change in that language as well.

There are facts which appear to bear out these predictions. Comparative evidence suggests that at some point in its history German
must have resembled Gothic and Icelandic in allowing disjoint reference and anaphor binding to relate nonsubject positions in an infinitive complement to the subject position of the clause containing
that complement. Such a stage is in fact attested in early German,
as illustrated in (20).

20.   diu künegin den$_j$ [$_S$ PRO$_j$ sich /(*sie) küssen] bat
      the queen$_i$ that-one   self$_i$ *her$_i$ to-kiss asked
      'The queen asked that one to kiss her.' (Wolfram,
                                                 Parzival 806:28)

As (21) shows, this is no longer possible in Modern German.[15]
Thus, German has changed in one of the expected ways.

(21)   Sie bat mich$_j$ [$_S$ PRO$_j$ sie /*sich zu besuchen]
       She$_i$ asked me   her$_i$ *self$_i$ to visit
       'She asked me to visit her.'

The comparative evidence of Gothic and Icelandic suggests further
that at some point in the history of German possessive pronouns like
ihre in (22) must have been interpreted as obligatory disjoint in
reference with the subject of the clauses containing them. Compare
(12) and (13a). As (22) shows, this is not the case in Modern German.

22.   Sie verkauften [$_{NP}$ ihre Bücher]
      they$_i$ sold   their$_{i,j}$ books

Thus, in German both the position of subject of NP and the position
of non-subject of infinitive complement have become opaque to obligatory disjoint reference with respect to constituent-external NPs,
as predicted by our hypothesis. Interestingly, German has also
undergone yet another change, which is not directly predicted by that
hypothesis. The comparative evidence of Old English, Gothic and
Icelandic indicates that the etymon of the possessive form seine in
(23) must once have been strictly reflexive--inadmissible in the
absence of a c-commanding antecedent. Again, this is no longer true
in German. Seine in (23) may either corefer or not corefer with the
subject er. That is, sein does not differ from ihr with respect to
antecedence requirements; the contrast between them is solely one
of gender and number.

23. Er$_i$ verkaufte [$_{NP}$ seine$_{i,j}$ Bucher]
     He$_i$ sold   his$_{i,j}$ books

This change in the status of sein, from anaphor to non-anaphor, does
not follow directly from our hypothesis; given (7), the position of
subject of NP is transparent to anaphor binding in the unmarked case,
so the change cannot be the direct result of adoption of the unmarked
definition of binding category. However, an explanation for this development is available within the framework of that hypothesis. We
have claimed, on the basis of comparative evidence, that NPs in
German did not originally constitute binding categories for pronouns
in NP subject position, but that they came to constitute binding
categories for such pronouns through the incorporation of the unmarked definition of binding category into German grammar. The
consequence of this change was that a pronoun could now be used as
subject of NP to express either intended noncoreference or intended
co-reference with the subject of the clause containing that NP.
Consider again example (22). Thus, the pronoun/reflexive opposition
ceased to be functional in the position in question. Under the reasonable assumption that languages tend not to maintain formal contrasts without functional value, we may expect that either one of the
forms will be abandoned as superfluous or they will be reinterpreted,
and come to participate in a different kind of opposition.[16] In
German the latter has taken place. Comparative evidence indicates
that the reflexive possessive sīn in Germanic could be used for all
genders and numbers. In German, however, sein has been limited to
masculine and neuter singular, and has thereby come into complete
complementarity with respect to these features with the surviving
pronominal possessives. Thus, the original pronominal/reflexives
opposition in NP subject position, rendered superfluous by a change
in the definition of binding catgories, has been replaced by an
opposition of gender and number.

   A similar development, with a different resolution, can be recon-
structed for the modern Romance languages. It is reasonable to suppose that at some point the distribution of pronouns and anaphors in
infinitive complements in pre-Spanish resembled the distribution
found in Latin sentences like (17a). (24) shows that this is no
longer the case; in Spanish, as in German, infinitive complements
have opaque to disjoint reference and anaphor binding. Again, this
development is consistent with our hypothesis, since it involves the
disappearance of a distribution which is characterized as marked
under that hypothesis.

24. Ellos me$_j$ ordenaron [$_S$ PRO$_j$ ayudarles/*ayudarse]
     They$_i$ me ordered   to-help-them$_i$ to-help-selves$_i$
     'They ordered me to help them.'

Significantly, Spanish, like German, has also neutralized the pro-
noun/reflexive distinction in NP subject position. Sus in (25),

historically a reflexive, is now used for both intended coreference
and intended noncoreference with the underscored subject position.
In Spanish, as in other Romance languages, the pronominal possessive
forms have been lost altogether. That is, these languages have uni-
formly eliminated the suus/eius distinction of Latin by discarding
the latter.

> 25. El hombre$_i$ habló con[$_{NP}$ sus$_{i,j}$ amigos]
>   The man$_i$ spoke with his$_{i,j}$ friends

It is clear that the explanation developed above for the neutralization of the reflexive/nonreflexive distinction in German can be extended to this development as well. The two cases differ only in
that Spanish and other Romance languages, rather than assigning new
values to the forms which participated in the now superfluous reflex-
ive/pronominal contrast in NP subject position, have simply abandoned
one of the two forms.

Returning to the problem of the English sentences in (9) and
(11), we see that it also admits a similar solution. Again, the
comparative evidence suggests that at some stage in pre-English, as
in pre-German, the position of subject of NP must have been transparent to disjoint reference. By Old English times it had become
opaque to disjoint reference, with the result that pronominal forms
in NP subject position could now be used to express intended coreference with the higher subject, as in (26).

> 26. (he) sealde [$_{NP}$ his hyrsted sweord]...ombihtþ egne
>   he$_i$ gave   his$_i$ decorated sword to-servant (Beow 672)

'He gave his decorated sword to a servant.'

Thus, the reflexive/nonreflexive opposition here had again become
superfluous. Rather than reinterpreting the forms, as in German, or
discarding the pronominal and generalizing the reflexive, as in
Romance, English chose the third available resolution: it abandoned
the reflexive form (a choice no doubt favored by the fact that the
earlier loss of the other Germanic reflexive forms had left the possessive reflexive without any paradigmatic support). For the same
reason, English did not subsequently introduce genitive forms of the
self-reflexives. They would have served no function for which the
genitive of he pronoun was not already adequate.

Thus, three indepedent historical developements--the change of
German sein from anaphor to nonanaphor, the abandonment of the
suus/eius distinction in Latin, and the gap in the English reflexive
paradigm--may all be derived as natural consequences of the imple-
mentation of the unmarked definition of binding category.

Notes

1. The definition of government assumed here is the following:

(i) β governs α iff β is an x^0, in the sense of the $\bar{X}$-Theory, and all maximal phrasal projections containing β also contain α and the converse.

2. The following definitions are assumed:
   (i). <u>SUBJECT</u> = a.) the AGR(EEMENT) element in finite clauses
         b.) the syntactic subject (i.e., [NP,S] or [NP,NP]) elsewhere

   (ii). γ is <u>accessible</u> to α iff
          a.) γ c-commands α
          b.) coindexation of γ , α would not violate the i-within-i condition of Chomsky (1982:212)

For detailed discussion, see Chomsky (1981).

3. The two differ in the following respects. First, (3a) predicts that (i) should be ungrammatical while (3b) predicts that it should be grammatical, since NP is a binding category for the anaphor in the sense of (3a), but not in the sense of (3b).

   (i). They found [$_{NP}$ some books [$_S$ for each other to read]]

My own judgement supports the latter prediction. Second, (3b) must be supplemented with a default condition in order to account for the obvious ungrammaticality of (ii).

   (ii). *[$_{\bar{S}}$ For each other to win] AGR$_i$ would be unfortunate

The AGR(EEMENT) element here is the only SUBJECT (in the sense of note 2) which c-commands the anaphor, and it is not accessible to that anaphor because of the i-within-i condition on accessibility. Thus, the anaphor lacks a binding category in the sense of (3b), and therefore to account for the ungrammaticality of (ii) it must be stipulated that root clauses are binding categories for elements which have no other binding category.

4. I assume here that the SUBJECT of an NP is its syntactic subject (NP,NP), in accordance with the definition given in note (2). Huang (1982) proposes an extended definition of SUBJECT under which the SUBJECT of an NP is the head N of that NP. For example, the SUBJECT of NP$_2$ in (4), so defined, would be <u>stories</u>. For the constructions under consideration in this paper this modification has no empirical effect. NP$_3$ is still not within the domain of an accessible SUBJECT within NP$_2$, given the i-within-i condition on accessibility. For discussion

of the cases in which the modification does have some empirical
consequences, see Huang (1982).

5. I adopt the general assumption that the principles of Universal
    Grammar are not absolute conditions on possible grammars, but
    that they characterize unmarked grammars. Thus, an individual
    grammar may produce constructions which do not conform to the
    requirements of these principles, but it is marked to the extent
    that it does so.

6. Alternatively, we might modify our assumptions about the nature
    of the forms in question. Chomsky (1981:217), for example,
    advances the possibility that in the intended interpretation
    forms like their in (5b) might be obligatory variants of the
    corresponding reflexives--i.e., that they are not strictly
    pronominal, but represent neutralizations of the pronoun/
    reflexive opposition. Left open, however, is the question of
    why such a neutralization should have taken place in this
    particular context and not in others.

7. The same result will follow if we take the SUBJECT of NP to be
    the nominal head of NP, as suggested in note (4). The NP in
    (5a) will still contain no SUBJECT accessible to each other's
    given the i-within-i condition on accessibility.

8. There apparently are cases in other languages in which the
    binding domain for reciprocals and the binding domain for
    reflexives do differ. Thráinsson (1979:296) observes that
    reciprocal binding in Icelandic is "clause bounded" while
    reflexive binding is not, as will be illustrated below.

9. Since these are extinct languages, ungrammaticality can only be
    inferred on the basis of nonattestation/avoidance of imitation
    of the model for translantion. The asterisk in (13a), for
    example, is justified by the fact the Gothic consistently uses
    the reflexive when coreference is intended, even though the
    Greek model for translation employs pronouns.

10. For example, we could assume that languages of the Icelandic
    type choose (3b) as the definition of binding categories for
    pronouns. This would correctly predict the disjoint reference
    facts of (12), for example. However, such as assumption would
    be inadequate since it would fail to capture certain other
    related differences in binding between languages of the English
    type and languages of the Icelandic type, as will seen below.

11. The similarity between Japanese and English involves only dis-
    joint reference assignment to pronouns. For discussion of the
    Japanese reflexive, see Oshima (1979).

12. Again, the asterisks are justified on the basis of nonattesta-
tion/avoidance of imitation of the model for translation. In
constructions like (17b), for example, Gothic systematically
exhibits a reflexive where coreference is intended, even though
the Greek model has a pronoun. There are a few exceptional
instances in which the Greek pronoun is translated by a Gothic
pronouns, e.g., Mk 1:10, Luk 1:73-74. See Harbert (1983a) for
an account of these cases.

13. Subjunctive complements in Icelandic are opaque to disjoint
reference but transparent to reflexive binding, as illustrated
in (i).

(i).   Jón segir að [$_S$María elski sig/hann]
       John says that Mary loves self$_i$ /him$_j$
       'John says that Mary loves him.' (Thráinsson (1976)

For discussion of these constructions, see Thráinsson (1976),
Maling (1981), Harbert (1983).

14. I am grateful to Alec Marantz for suggesting this possibilty to
me.

15. With a small set of verbs taking A.c.I complements in Modern
German, e.g., the causative lassen, reflexives in the complement
may be bound to the higher subject under restricted
circumstances, as in (i).

(i).   Der Chef lässt [die Leute für sich/ihn arbeiten]
       The boss has the people for self$_i$/him$_i$ work
       (Reiss 1976:27)

For a discussion of these cases, see Harbert (1983).

16. As illustrated by the Chinese examples in (6), however, it is
also possible for a language to maintain the pronoun/reflexive
distinction even when it is not required to by restrictions on
possibilities for coreference.

REFERENCES

Andrews, A. 1976. The VP Complement Analysis in Modern Icelandic.
    Papers from the Sixth Annual Meeting, North Eastern Linguistic
    Society 1-22.

Chomsky, N. 1981. Lectures on Government and Binding. Dordrecht:
    Foris.

Harbert, W. 1983. (to appear) On the Definition of Binding Domains.
    Proceedings of the Second West Coast Conference on Formal Lin-
    guistics. Stanford.

_____1983a. (to appear) Germanic Reflexives and the Im-
    plementation of Binding Conditions. In I. Rauch and G. Carr,
    eds., Language Change. Indiana University Press.

Huang, C.-T.J. 1982. Logical Relations in Chinese and the Theory of
    Grammar. Doctoral Dissertation, MIT.

Maling, J. 1981. Non-Clause-Bounded Reflexives in Icelandic. In
    T. Fretheim and L. Hellan, eds., Papers from the Sixth Scandin-
    avian Conference on Linguistics.

Oshima, S. 1979. Conditions on Rules: Anaphora in Japanese. In
    G. Gedell, E. Kobayashi and M. Muraki, eds., Explorations in
    Linguistics: Paper in Honor of Kazuko Inoue. Tokyo: Kenkyusha.

Reis, M. 1976. Reflexivierung in deutschen A.c.I.-Konstruktionen.
    Ein transformationsgrammatisches Dilemma. Papiere zur
    Linguistik 9:5-82.

Thrainsson, H. 1976. Reflexives and Subjunctives in Icelandic.
    Papers from the Sixth Annual Meeting, North Eastern Linguistic
    Society. 225-239.

_____1976a. Some Arguments Against the Interpretive Theory of
    Pronouns and Reflexives, Harvard Studies in Syntax and Semantics
    2.573-624.

_____1979. On Complementation in Icelandic. New York:
    Garland Publishing.

MARKEDNESS AND THE ZERO-DERIVED DENOMINAL VERB IN ENGLISH:

SYNCHRONIC, DIACHRONIC, AND ACQUISITION CORRELATES

Terence Odlin

Ohio State University

Otto Jespersen (1938) called zero-derivation a 'characteristic peculiarity' of English. The basis for his remark seems to be that English makes more extensive use of zero-derivation patterns than do other European languages. Moreover, verbs derived from concrete nouns, hereafter referred to as CONCRETE DENOMINAL VERBS, as in

1) The sailor knifed the bandit.

seem to be especially productive in English: Clark and Clark (1979) have found over 1300 attested cases of such verbs. Yet it is not only their productivity which is unusual. The purpose of this paper is to show that concrete denominal verbs are peculiar even with respect to another type of zero-derivation in English, DEADJECTIVAL VERBS, as in

2) The wind at the top of the mountain cooled us.

The extra peculiarity of concrete denominal verbs has interesting implications for a notional theory of parts of speech and supports the treatment of markedness as a scalar construct.

Determining the criteria for zero-derivation is naturally more problematic than with overtly marked forms where one can assume that the form having an extra morpheme is the derived form. In fact, during the heyday of the American Structuralist movement, the notion of zero-derivation was not a commonplace in morphological analysis. In the case of denominal verbs, there was a reluctance to see them as indeed derived (e.g. in the analysis of Nida 1948).

Since the Structuralist period, however, derivational relationships have often been posited. Marchand (1974) proposes several
criteria to define zero-derivations. Some, such as stress, are less
important in his analysis, and will not be discussed here. The most
important criterion is what Marchand calls 'semantic dependence',
which he sees "...as often as not sufficient in itself to solve the
question of derivational relationship" (1974:242). He explains
semantic dependence thus:

> ...the content analysis of the verb must necessarily include
> features of the substantive...The verb knife is naturally
> analysable as "wound with a knife" whereas the substantive knife
> does not lean on any content features of the verb knife...
> (1974:245)

Another criterion often useful is what Marchand calls "restriction
of usage". Although it can involve other characteristics, it often
is a question of frequency. For example, "The substantives father
and author...are high frequency words whereas the verbs are used
much less frequently" (1974:246). Another important criterion
Marchand terms 'semantic range' and provides the following example:

> The content of the substantive butcher appears also in the verb
> butcher, but with restricting depreciative features, involving a
> particular field of reference. The substantive enters into the
> verb with a narrow specification of its content, butcher being
> emotionally motivated as "slaughter of animals". (1974:247)

To compare zero-derived denominals and zero-derived deadjectivals
might thus seem to involve no more than a question of comparative
measures of the semantic dependence, the restriction of usage, and
the semantic range of the items in question. Yet while Marchand's
criteria are probably about the best which can be formulated, it is
not so simple using them in a comparison of the two types of zero-
derivations. For one thing, semantic dependence does not seem to be
any more characteristic of one type of zero-derivation than of another. Thus, where knife as a verb means 'cut or stab with a knife'
and cool means 'cause to become cool', both verbs evidence an equal
degree of semantic dependence. Thus, measurement of comparative
semantic dependence is a dubious undertaking.

Given that elaborate word frequency lists such as the Lorge and
Thorndike (1938) semantic word count exist, it might seem feasible
to invoke the restriction-of-usage criterion by taking the frequen-
cies of cool as a verb and comparing them with knife as a verb. Yet
here there are also problems. For example, should metaphoric uses
be included in the frequency count? If one decides to count them,
should they be included together with literal occurrences or separately? Choosing the one course or the other might seriously bias
the count. Yet if one decides not to count them, one then has the
problem of determining what is or is not metaphoric, a problem

which Lakoff and Johnson (1980) have suggested may often be insol-
uble. For instance, is the deadjectival verb in

3)   Jack's face was crimson with anger, but then he cooled down.

metaphoric or not?

Another possible measure might seem to be semantic range. Here,
however, the problem of metaphors is even more intractable: if one
decides to count metaphoric extensions, the number of conceivable
senses is not necessarily restricted to even the copious listings of
the Oxford English Dictionary. Moreover, it could well be that there
are an infinite number of metaphoric sense extensions. There is also
the question of whether denominal and deadjectival sense extensions
are really comparable. The most important sense of knife which is
glossed in the OED is 'To use a knife' and the most important (transitive) sense of cool is 'To make cool'. Such glosses, however,
imply an equal semantic range in the two verbs, a questionable
implication as subsequent analysis will show.

While the measures just suggested are inadequate, there are
others which can be usefully explored. One is to compare the range
of derivational options (including zero) available with denominal
verbs and deadjectival verbs. As Marchand (1969) notes, zero is
generally the only option available to derive a verb from a concrete
noun, although there are a few exceptions such as with the prefix in
enslave. In contrast, verbs may be derived from adjectives in several ways: by means of zero, by prefixes as in endear and renew,
and by several suffixes as in deepen, formalize, purify, and liqui-
date. With such a range of derivational markers available, it is
not surprising that zero-derived deadjectival verbs are not as productive a subclass as are zero-derived denominal verbs. However,
upon looking at a list of the most frequent words in English (e.g.
West 1953), it is evident that while many concrete nouns routinely
undergo zero-derivation, many do not derive as verbs either by zero
or by an overt marker, except in noun usages. In contrast, many
more adjective root morphemes routinely undergo some kind of derivation into verbs, whether by zero or by other means.[1]

One way to compare the semantic range of denominal and deadjectival verbs that are zero-derived is to look at the co-occurrence
restrictions such as those on direct objects in the case of transitive verbs. While knives can be used to cut almost anything, the
verb knife generally takes animate and usually human patients.
Compare sentence 1 with

4)   ?The sailor knifed the dog.
5)   ??The sailor knifed the roast beef.[2]

In contrast, zero-derived deadjectivals such as cool can often take animate or inanimate patients.[3] Compare sentence 2 with

  6)   The wind at the top of the mountain cooled the beer.

Moreover, denominal verbs tend to carry much more detailed denotations of particular actions: that is, we get much more information
as to where or how the action took place in a concrete denominal as
opposed to a deadjectival verb. For example, sentences such as

  7)   The attendant bagged the groceries.

  8)   The bandit was knifed with a switchblade.

preclude sentences such as

  9)   ??The attendant bagged the groceries in a box.

  10)   ??The bandit was knifed with a revolver.

In contrast, stative verbs tend to be much less specific as to
manner.

In sentence 11

  11)   We cooled the beer.

there are few exclusions as to the place or manner in which the action was performed. Thus any of the following sentences are
possible:

  12)   We cooled the beer in the fridge.

  13)   We cooled the beer in the ice chest.

  14)   We cooled the beer on New Year's Eve by leaving it out on
        the roof

  15)   We cooled the beer by leaving it on top of the mountain
        overnight.

  16)   We cooled the beer by lowering it into the Mariana Trench.

Thus, while on the basis of dictionary glosses denominal and deadjectival verbs might seem to have roughly equivalent semantic ranges,
the former are generally more restricted.

Although diachronic facts must not be confused with synchronic ones,
there is a high correlation between synchronic and diachronic derivation. In the case of zero-derivation, we might expect that a
synchronically derived form will not be attested as early in the lan-

guage as will the base form. However, this expectation turns out to
be much truer of denominal verbs than of deadjectival ones.
Tables 1 and 2 show a remarkable difference between the two types in
terms of the earliest attestation dates for base and derived forms
as given in the OED.[4]

In the case of 30 denominal verbs, which were randomly sampled
from West (1953), 21 of the attestations are over a hundred years
after the earliest attestation of the noun base form. In contrast,
only 11 of the randomly sampled deadjectival verb attestations are
over a hundred years old. In almost two-thirds of the cases, the
verb form is hardly any more recent than the adjective base form,
and in some cases it is considerably earlier.

Attestation dates do not, of course, necessarily mean that the
derived forms really are more recent additions to the language. Yet
if that interpretation is excluded, the only other which seems admis-
sible is that the derived form was, at least in its earliest years
of use, less frequently employed than the base form. The latter
interpretation, of course, indirectly implies Marchand's restriction-

Table 1. Thirty concrete noun lexemes which zero-derive into verbs.
Dates are earliest attestations in the OED for comparable senses of
each lexeme which have continued in use to the present.

Lexeme	Noun	Verb	Difference
BAG	1230	1573	343
BARREL	1305	1624	319
BASKET	1300	1583	283
BICYCLE	1868	1876	8
BOAT	891	1673	782
BOTTLE	1375	1641	266
BRIDGE	1000	1000	0
BRUSH	1377	1460	83
CAGE	1225	1577	352
CHAIN	1300	1393	93
CLOCK	1371	1883	512
COLLAR	1300	1601	301
CORK	1530	1659	129
CROWN	1085	1175	90
FACE	1290	1632	342
FISH	825	888	63
FORK	1000	1802	802
KNIFE	1175	1865	690
PEN	1300	1490	190
PIN	1100	1377	277
PLOUGH	1200	1460	260
ROOT	1150	1440	290
SHELL	1375	1694	319
SHOE	950	897	-57
SOAP	1000	1585	585
STONE	888	1200	312
TELEPHONE	1866	1880	14
TREE	825	1781	956
WING	1175	1611	436
WHIP	1340	1386	46

of-usage criterion. Thus whether we view attestation dates as a
measure of how recent a zero-derivation is in the language, or whether we view them as a measure of the frequency of the zero-derivation at some earlier period in the history of the language, once
again denominal verbs are very different from deadjectival verbs.

Aside from synchronic and diachronic evidence, the peculiarity
of concrete denominal verbs is also evident in second language acquisition. Sentences 17 to 20 are examples of a type of second language error which can be termed PART-OF-SPEECH ANOMALIES:

17)   The exploration of the BENEATHS of the sea is almost
       complete.

18)   In front of the altar, we bow HUMILITY and pray ritually...

19)   ...this situation will find a PEACEFULLY solution...

20)   When he talked about Chairman's Mao's death, he tried to
       keep from TEARING but failed.[5]

Table 2. Thirty adjective lexemes which zero-derive into verbs.
Dates are earliest attestations in the OED for comparable senses of
each lexeme which have continued in use up to the present.

Lexeme	Adjective	Verb	Difference
BUSY	1225	1000	-225
BLIND	1000	1300	300
CALM	1400	1399	-1
CLEAR	1297	1374	77
COMPLETE	1380	1530	150
CONTENT	1400	1477	77
COOL	1000	1000	0
CORRECT	1460	1340	-120
DRY	1000	888	-112
EMPTY	971	1000	129
EXACT	1533	1529	-4
FAINT	1320	1400	80
FIRM	1374	1374	0
FREE	888	1000	112
FREQUENT	1531	1555	24
HOLLOW	1205	1450	245
IDLE	950	1652	702
LOOSE	1225	1225	0
NARROW	893	1000	117
OPEN	888	1000	112
PALE	1300	1374	74
PROMPT	1432	1340	-92
QUIET	1375	1550	175
READY	1200	1350	150
SINGLE	1340	1575	235
SLOW	1340	1557	217
SMOOTH	1050	1440	390
STEADY	1574	1530	-44
WARM	888	900	12
WRONG	1300	1330	30

In the past four years, I have compiled over three hundred examples of such anomalies from the speech and writing of over one hundred EFL students with various first languages. While most such errors are not particularly frequent, sentence 20 is an instance of a subclass which is especially rare: i.e. concrete nouns ungrammatically derived as denominal verbs. I have found only six instances of that subclass--and most of those were rather marginal instances. Thus, either in terms of the total number of anomalies (6 out of 300), or in terms of the number of students who produced them (6 out of 100 individuals), concrete denominal verb anomalies are exceedingly rare. At the same time, other types of anomalies were not so uncommon. For example, denominal verb anomalies involving overt derivational markers as in

21)   ...efforts to MANAGEMENT Chinese society...

and anomalies which were zero-derived but where the base noun had an abstract denotation as in

22)   Japan have SUCCESSED in renewal...

were much more common.[7] This is strange for several reasons. First, concrete nouns are, after all, among the first items within a single part-of-speech class learned in large numbers. Nevertheless, errors occur extremely rarely. Second, concrete denominal verbs are very productive in English, as Clark and Clark (1979) have shown. Third, zero-derived verbs are much more commonly based on concrete nouns than on abstract ones and derivations of overtly marked nouns into verbs are also rare, according to Marchand (1969).

The peculiarity of concrete denominal verbs in second language acquisition is also seen in elicitation data (Odlin 1983). In a grammaticality judgment test given to eight different groups of EFL students, sentences with denominal verbs were generally rated as less grammatical. Thus, as seen in Tables 3 and 4, sentences with verbs such as bottle and lamp were rated lower than sentences with stative shifts such as cleared and handsomed.[8] In contrast, native speakers tended to show no preference for stative shifts over denominal verbs per se, though grammaticality was definitely a factor in their ratings. Thus they tended to rate sentences with verbs such as lamp and handsomed lower than ones with verbs such as bottle and cleared.

These generally negative ratings of denominal verb sentences were given primarily by speakers of Spanish, but similar results appear with a small group of Vietnamese speakers, as seen in Table 4. Thus the test results tend to support the production data which indicate that denominal verb anomalies are an extremely rare type of error in EFL. Test performances cannot be attributed simply to the possibility that learners rate denominal verb sentences lower because

Table 3. Classification of items on grammaticality judgment test.
CS = concrete shift, SS = stative shift, C = grammatical, i =
ungrammatical. The words given in upper case letters here were in
lower case on the test.

(cCS: 10 items) Mr. Garza wants to BOTTLE the wine that he
                  made.

(iCS: 10 items) It was dark so policemen had to LAMP the room.

(CSS: 10 items) Before the test the students CLEARED their
                  desks.

(iSS: 10 items) In ten minutes the barber HANDSOMED George's
                  hair.

they may have never heard <u>bottle</u> or other lexemes employed as denominal verbs. Among the various groups, there was a pronounced tendency to rate ungrammatical sentences with denominal verbs such as
<u>lamp</u> lower than iSS sentences, even though EFL students could hardly
have heard either type of sentence before. While previous exposure
probably plays a role in some groups' performances, there should not
be such consistently lower ratings of denominal verbs if only previous exposure were involved.[9]

     From all of the preceding evidence, it is clear that both con-
crete denominal verbs and deadjectival verbs are subject to constraints which can be seen as indicators of markedness. This is
especially true of concrete denominal verbs. Although criteria such
as neutralization are not applicable, there is a clear parallelism
between denominal verbs and examples of marked items such as spatial
terms with negative polarity studied by Clark (1969). Lexemes such
as <u>low</u> and <u>small</u> are morphologically restricted in that they rarely
take on nominal forms as in <u>lowness</u> and <u>smallness</u>. Likewise there
are morphological restrictions on denominal verbs: i.e. most are
derived only with zero. There are also semantic restrictions on
spatial terms with negative polarity: for example, <u>Jeff is as short</u>

Table 4. Mean ratings for each type on test by the group of Vietnamese speakers (n=5). The difference between CS and SS ratings was
statistically significant (p <.011). Standard deviations are given
in parentheses.

	VT
a) cCS (e.g. BOTTLE)	2.2 (.6)
b) iCS (e.g. LAMP)	1.7 (.7)
c) cSS (e.g. CLEARED)	3.1 (.3)
d) iSS (e.g. HANDSOMED)	2.2 (.5)

as George presupposes that both individuals are short whereas the analogous sentence that has a spatial term with positive polarity (i.e. Jeff is as tall as George) carries no presupposition that the individuals are tall. In a somewhat different way, denominal verbs are also semantically restricted: in comparison with deadjectival verbs, they generally have more restrictions on what direct objects they can take when they are transitive, and they are generally more specific as to action and manner. Historically they are either less frequent in the early years of their use, or they lag historically in terms of being used at all. In second language acquisition, over-generalized concrete denominal verbs are much rarer than other part-of-speech anomalies, and on a grammaticality judgment test, concrete denominal verbs tend to be rated as less grammatical than sentences with stative shifts.

Bolinger (1975) sees any semantic extension over time as a kind of zero-derivation. Conversely, any zero-derivation can be seen as a kind of semantic extension. Why then should the semantic extensions seen in concrete denominal verbs be so restricted? The OED editors provide an interesting insight in their preface to the dictionary:

> Some words have only one invariable signification; but most words that have been used for any length of time in a language have acquired a long and sometimes intricate series of signi-fications, as the primitive sense has been gradually extended to include allied or associated ideas, or transferred boldly to figurative and analogical uses. This happens to a greater extent with relational words, as prepositions...than with notional words, as verbs and nouns; of these, also it affects verbs and adjectives more than substantives; of substantives, it influences those which express actions, qualities, and mental conceptions...more than those which name, and are, as it were, fixed to material objects. (1933:xxxi)

The editors point out, subsequently, that concrete nouns also have a capacity to acquire a different sense, but their point is clear that diachronically it is feasible to see concrete nouns as notionally very restricted.

A more recent proponent of a notional theory of parts of speech, Lyons (1977), has suggested that the reason for the relatively more predictable grammatical behavior of concrete nouns lies in their on-tological determinacy: i.e. for whatever variations there may be between speakers of different languages, people generally perceive and categorize physical objects in the same way and this has a reflex in the lexical codings of nouns. In contrast, categorization of more abstract realia such as events, attributes, and places is more variable, and this is reflected in the greater variability across languages in terms of verb, adjective, and adverb systems—

including the fact that in some languages, as in the Sino-Tibetan
family, there is little basis for positing a separate lexical category of adjectives. Lyons is careful to qualify his account of what
criteria are relevant to the psychological reality of a category:
some members of the noun category are taken to be the best examples
of the category, some verbs of the verb category, and so forth.[10]
By his own avowal, this approach is similar to the analysis of color
terminology by Berlin and Kay (1969). It also has affinities with
the work of some Generative Semanticists (e.g. Lakoff 1972 and Ross
1973) and even more with the prototype theories of Rosch (1978) and
others. (Since Lyons, Bates and MacWhinney (1982) have elaborated a
discussion of part-of-speech theory along similar lines.) Such an
approach has the dual advantage of facilitating the exploration of
the psychological reality of categories in general--including parts
of speech--and in facilitating understanding the often-noted tendency in natural languages to show at least a basic noun-verb distinction.[11]

        In addition to the observation of the OED editors, the facts pre-
sented in this paper about denominal verbs suggest that while the
semantic extensions of concrete nouns into verbs are a productive
lexical pattern in English, they are nevertheless constrained in
many important ways in comparison with deadjectival verbs. If concrete denominal verbs and deadjectival nouns were compared, similar
constraints would be evident, the constraints generally being compatible with a notional theory of parts of speech. There is thus an
implication in all this for theories of markedness. Discussions of
a general theory of markedness have, as Lyons (1977) observed, tended
to treat markedness as a matter of kind rather than as a matter of
degree. Moreover, much of the work on markedness in morphology has
naturally put great emphasis on the role of overt markers, an emphasis very conducive to a binary characterization of markedness (e.g.
Jakobson 1964). However, a binary characterization of zero-derivations in English seems neither necessary nor desirable. Marchand's
criteria of semantic dependence, restriction of usage, and semantic
range resemble several common criteria to determine markedness.
Thus, both concrete denominal verbs and deadjectival verbs can be
seen as cases of markedness despite the absence of overt markers.
Yet at the same time, the comparison of two types of zero-derived
verbs in this paper suggests that concrete denominal verbs show more
markedness than do deadjectivals. Thus an adequate theory should
treat markedness as a scalar construct. Markedness is a notion
pregnant with interesting possibilities for the understanding of
language, but it should not be regarded as pregnancy is--as a matter
of all or nothing.

NOTES

        [1]The problem of nonce usage impedes giving any precise estimate

of the relative productivity of any morphological process. In a recent study of the semantics of particular verbs and their derived
adjectives, Brown and Fish (1983) argue that what is productive in
the derivational morphology of a language is closely connected to
the expectations that people have about the world and especially
about each other. On the basis of experimental and other kinds of
evidence, they claim that '...no firm line can be drawn between
actual words and possible words' (1983:246). Accordingly, determining what is or is not productive might sometimes require making
dubious assumptions about the usualness of a given state of affairs.
Nevertheless, at least half of the adjectives in the word frequency
count of West (1953) seem to commonly undergo derivation into verbs,
whether by overt or by zero marking, whereas a much smaller proportion of the nouns (of those which can be reliably called 'concrete')
commonly undergo derivation into verbs by any means.

[2]In other cases, the restrictions work in different ways. For
example, shoe is normally a verb requiring a hoofed animal for direct object, and bottle implies putting liquids in a container as
part of a manufacturing process. Thus even though the sentences
Mrs. Wilson shoed her son and The children went to the playground to
bottle some sand are understandable, they are not very plausible
utterances.

[3]There are, of course, some semantic restrictions on a number
of the deadjectivals. Blind, for example, can only take animate
patients. However, in this and several other cases, the semantic
restriction is not arbitrary, as it is in the case of knife and many
other denominal verbs. Only animate beings can he blind or can be
blinded, whereas it is possible to use a knife to cut inanimate as
well as animate patients, even though this possibility is not
realized in the denominal verb.

[4]The following criteria were used in selecting the earliest
citation dates in the OED: 1) the citation had to be of a lexeme
used in a sentence, not in a dictionary entry or a textual gloss;
2) participial and gerundive constructions were not counted; 3)
only denotations of concrete objects which would correspond to the
notion of 'basic level objects' (Rosch et al. 1976, Rosch 1978)
were considered; 4) the sense of the noun and verb citations for
each lexeme had to be comparable and still in use in the language--
thus, for example, tree has a citation from 1650 with the meaning
'grow into a tree' which, according to the OED, is now obsolete--
thus a later citation (from 1781) is counted as the earliest one;
5) following the analysis of Clark and Clark (1979), non-figurative
senses of denominal verbs--insofar as these could be reliably identi-
fied--were considered as the sense most comparable to the noun
senses. [5]The student who wrote this said that he intended tear as
a synonym of cry. Of course it is possible for tear to be a verb as
in The doctor teared my eyes before she examined them. Yet given

that any part-of-speech anomaly is a semantic extension, it would
not be plausible to argue that the student's use of ...keep himself
from TEARING is grammatical unless most native speakers would use
tear in such a sentence. The problem is the same as in examples
which Clark (1982) found of children who use verbs that lexemically
are identical with ones used by adults but that are semantically anomalous, as in You have to SCALE it [some cheese] first. Yet even if
anyone considers the students' use of tear to be more acceptable
than the child's use of scale, the claim made in this paper about
the infrequency of concrete denominal verb anomalies in second
language acquisition is strengthened instead of weakened. In fact,
virtually all of the anomalies involving such verbs could, for one
reason or another, be considered a marginal case.

[6]The general infrequency of part-of-speech anomalies has been
noted in first language acquisition studies (e.g. Slobin 1973), and
this has led to renewed discussion of theories of parts-of-speech
and the significance of such theories for language acquisition research (cf. Maratsos and Chalkley 1980, Bates and MacWhinney 1982).

[7]In one controlled comparison of the relative frequency of the
two types of errors, Odlin (1983) found abstract denominal verbs to
be five times more frequent than ones involving concrete denominal
verbs.

[8]On the DN test there were different subcategories of SS sentences, but as there were no significant differences in performances
attributable to these subcategories, there is no discussion of them
here. (cf. Odlin 1983)

[9]In fact, the analysis-of-variance procedure used shows that
the group of Vietnamese speakers, as well as one other group, could
apparently not distinguish to any significant degree between sentences they might have heard before (i.e. c sentences) from ones
they could not have heard before (i.e. i sentences). Moreover,
there are other reasons to doubt the importance of previous exposure
as a factor. (cf. Odlin 1983)

[10]Lyons (1977) sees a need to distinguish between two possible
types of universals in a theory of parts of speech: lexical and
functional universals. In fact, he finds that analyses of languages
such as Nootka (e.g. Swadesh 1938) may compel a denial of any real
lexical universals (e.g. nouns and verbs in all languages), and
allow only for functional ones (e.g. nominal and verbal constructions
in all languages). However, the Structuralist analyses of Nootka
and other languages that deny any distinct lexical status for nouns
and verbs have been challenged, notably in the work of Dixon (1977).
Thus it may be possible to modify Lyons' position to allow for both
lexical and functional universals in a theory of parts of speech,
and the present discussion of denominal verbs assumes just such a
possibility.

One issue of considerable importance is, of course, the extent
to which zero-derived concrete denominal verbs appear in other languages. Since the discussion of zero-derivation in morphology has,
in most studies, been raised in the context of European languages
(most notably English), it is especially important to see if non-
Indo-European languages that use zero-derivation make much use of
concrete denominal verbs. If they do, it would be interesting to
see if the findings on English verbs on semantic restrictions, diachrony, etc. hold true. In the descriptions of zero-derivation in
non-Indo-European languages that I have seen (e.g. Le 1948, Chao
1968, Truong 1979, Rose 1973, Chapin 1978) the information is generally inconclusive and at times contradictory.

REFERENCES

Bates, Elizabeth, and Brian MacWhinney. 1982. Functionalist appro-
    ches to grammar. In Lila Gleitman and Eric Wanner, eds.
    Language Acquisition: The State of the Art, 173-218. Cambridge:
    Cambridge University Press.
Berlin, Brent, and Paul Kay. 1969. Basic Color Terms. Berkeley:
    University of California Press.
Bolinger, Dwight. 1975. Aspects of Language. New York: Harcourt,
    Brace and Jovanovich.
Brown, Roger, and Deborah Fish. 1983. The psychological causality
    implicit in language. Cognition 14.237-273.
Chao, Yuen R. 1968. A Grammar of Spoken Chinese. Berkeley:
    University of California Press.
Chapin, Paul. 1978. The Language of Easter Island: a characteristic
    VSO language. Winfred Lehmann ed., Syntactic Typology,
    139-68. Austin: University of Texas Press.
Clark, Eve. 1982. The young word-maker: a case study of innovation
    in the child's lexicon. In Lila Gleitman and Eric Wanner, eds.,
    Language Acquisition: The State of the Art, eds. 390-426.
    Cambridge: Cambridge University Press.
Clark, Herbert. 1969. Linguistic processes in deductive reasoning.
    Psychological Review 76.387-404.
Clark, Eve, and Herbert Clark. 1979. When nouns surface as verbs.
    Language 55.767-811.
Dixon, R. M. W. 1977. Where have all the adjectives gone? Studies
    in Language 1.19-80.
Jakobson, Roman. 1964. Zur Struktur des russischen Verbums. In
    Josef Vachek, ed. A Prague School Reader in Linguistics, 237-59.
    Bloomington: University of Indiana Press.
Jespersen, Otto. 1938. Growth and Structure of the English Language.
    New York: Doubleday.
Lakoff, George. 1972. Hedges: a study in meaning criteria and the
    logic of fuzzy concepts. In Paul Peranteau, Judith Levi, and
    Gloria Phares, eds. Papers from the 8th Regional Meeting,
    Chicago Linguistic Society, 183-228. Chicago: Chicago
    Linguistic Society.

Lakoff, George and Mark Johnson. 1980. Metaphors We Live By. Chicago: University of Chicago Press.

Le Van Ly. 1948. Le parler vietnamien. Paris: Huong Anh.

Lorge, Irving, and Edward Thorndike. 1938. A semantic count of English words. Teachers College, Columbia University.

Lyons, John. 1977. Semantics. Cambridge: Cambridge Unversity Press.

Maratsos, Michael, and Mary Ann Chalkley. 1980. The internal language of children's syntax: the ontogenesis and representation of syntactic categories, 127-215. In Keith Nelson, ed. Children's Language, New York: Gardner Press.

Marchand, Hans. 1969. The Categories and Types of Present-Day English Word-Formation. Munich: Oscar Beck.

Marchand, Hans. 1974. A set of criteria for the establishing of derivational morphemes. In Dieter Kastovsky, ed. Studies in Syntax and Word-Formation, ed. by 242-52. Munich: Wilhelm Fink.

Nida, Eugene. 1948. The identification of morphemes. Language 24.414-41.

Odlin, Terence. 1983. Part-of-speech anomalies in a second language. Unpublished dissertation, University of Texas at Austin.

Oxford English Dictionary on Historical Principles. 1933. Edited by J. A. H. Murray, H. Bradley, W. A. Craigie, and C. T. Onions. Oxford: Clarendon Press.

Rosch, Eleanor. 1978. Principles of categorization. In Eleanor Rosch and Barbara Lloyd, eds. Cognition and Categorization, 24-48. Hillsdale, N. J.: Lawrence Erlbaum.

Rosch, Eleanor, Carolyn Mervis, Wayne Gray, David Johnson, and Penny Boyes-Braem. 1976. Basic objects in natural categories. Cognitive Psychology 8.382-439.

Rose, James. 1973. Principled limitations on productivity in denominal verbs. Foundations of Language 10. 509-26.

Ross, John. 1973. A fake NP squish. In Charles-James Bailey and Roger Shuy, eds. New Ways of Analyzing Variation in English, 96-140. Washington, D. C.: Georgetown University Press.

Slobin, Dan. 1973. Cognitive prerequisites for the development, In Charles Ferguson and Dan Slobin, Studies in Child Language Development 175-208. New York: Holt, Rinehart, and Winston.

Swadesh, Morris. 1938. Nootka internal syntax. International Journal of American Linguistics 11.77-102.

Truong Van Chinh. 1970. Structure de la langue vietnamienne. Paris: Imprimerie Nationale.

West, Michael. 1953. A General Service List of English Words. London: Longmans, Green and Co.

THE SEMIOTIC THEORY OF ERGATIVITY AND MARKEDNESS

Sebastian Shaumyan

Yale University

1.   The Correspondence Hypothesis

The theories of ergativity now prevailing claim that the syntactic organization of ergative constructions in ergative languages
follows the pattern of accusative constructions in accusative languages.

This claim is based on a crude analysis which disregards both
the meaning and morphological form of syntactic units. Language is
a sign-system; therefore, linguistic units are bilateral: any linguistic unit consists of a sign and its meaning which presuppose
each other. Since syntactic units, as any linguistic units, are
bilateral, syntactic analysis cannot be separated from the analysis
of the meaning and the morphological form of the syntactic units.

In contrast to the thesis that ergative constructions are really
accusative in their syntax and ergative only in their morphology, a
hypothesis is proposed that the syntax of ergative constructions is
a counterpart of their morphology and that ergative constructions
are different from accusative constructions both in morphology and
syntax. This hypothesis I call the Hypothesis of The Correspondence
of Morphology and Syntax in Ergative Constructions (henceforth the
Correspondence Hypothesis).

As is well known, some ergative languages, like Chukchee, have
both ergative and accusative morphologies, that is they are not one
hundred percent ergative, but ergative only to a certain degree.
The notion 'degree of ergativity' suggested by Comrie (1978) raises
the important problem of determining and explaining the partial
identity of the syntactic organization of some ergative languages

169

with the syntactic organization of accusative languages. We must
strictly distinguish two entirely different questions. One question is whether the syntactic structure of the ergative construction is identical with or different from the syntactic structure of
the accusative construction. This question is completely independent
of another one, namely, whether a given language uses only ergative
constructions or also accusative constructions, that is, whether or
not its syntactic organization is partly identical with the syntactic organization of accusative languages.

The Correspondence Hypothesis is a consequence of the <u>Principle
of Semiotic Relevance</u>:

> Only those distinctions between meanings are relevant which
> correlate with the distinctions between their signs, and
> vice versa, only those distinctions between signs are rele-
> vant which correlate with the distinctions between their
> meanings.[1]

The Principle of Semiotic Relevance characterizes the properties of two basic linguistic relations:

1. <u>Sign of</u>: X is a <u>sign of</u> Y

2. <u>Meaning of</u>: Y is a <u>meaning of</u> X

Speaking of signs, we mean a binary relation <u>sign of</u>. Speaking
of meaning, we mean a binary relation <u>meaning of</u>.

<u>Sign of</u>. X is a sign of Y if X means Y, that is if X carries
the information Y. For instance, the sound sequence /bɛd/ carries
information 'bed', i.e. it means 'bed'; therefore, /bɛd/ is a sign
of 'bed'.

A sign is not necessarily a sound sequence. It may be the
change of stress (compare <u>cónvict</u> and <u>convíct</u>), an alternation
(compare <u>take</u> and <u>took</u>), the change of word order (compare <u>the
hunter killed the bear</u> and <u>the bear killed the hunter</u>), or the
change of a context (compare <u>I love</u> and <u>my love</u>). There may be a
zero sign: for example, if we compare <u>long, longer</u>, and <u>longest</u>, we
see that <u>er</u> is a sign of the comparative degree and <u>est</u> is a sign of
the superlative degree, but the positive degree is expressed by the
absence of any sound sequence with <u>long</u>, that is by a zero sign.

The opposition <u>sign:meaning</u> is relative. There may be an inter-
change between these entities. For example, the letter <u>p</u> in the
English alphabet normally denotes the sound /p/. But when we refer
to the English letter <u>p</u>, we use the sound /p/ as a name, that is as
a sign of this letter. Further, the meaning of a sign may serve as
a sign of another meaning. Thus, <u>lion</u> is a sign of a large, strong,

flesh-eating animal. This meaning of the sign <u>lion</u> can be used as a sign of a person whose company is very much desired at social gatherings, for example, a famous author or musician.

It follows from the foregoing that the proposed concept of the sign is considerably broader than the common concept of the sign.

<u>Meaning of</u>. The proposed concept of meaning is much broader than the traditional understanding of the term 'meaning.' My concept of meaning covers all kinds of information including various grammatical relations. As was shown above, the notion of meaning is relative: a meaning may be a sign of another meaning.[2]

Here are some examples to illustrate the Principle of Semiotic Relevance. The English word <u>wash</u> has different meanings in the context of the expressions <u>wash hands</u> and <u>wash linen</u>. But the distinction between these two meanings is irrelevant for the English language because this distinction does not correlate with a distinction between two different sound sequences: in both cases we have the same sound sequence denoted by <u>wash</u>. Therefore, these two meanings must be regarded not as different meanings but as two variants of the same meaning. On the other hand, the meaning of the Russian word <u>myt'</u> and <u>stirat'</u> correlates with the distinction between different sequences of sounds and therefore is relevant for Russian language. A similar reasoning applies to other examples. As to the relevant and irrelevant distinctions between signs, consider, for instance, substitution of [ə] for [I] in the penultimate syllables of terminations such as <u>-ity</u> or <u>-ily</u>: [ə'bIləti] for [ə'bIlIti] (ability). Since the distinction between the two signs is not correlated with a distinction between their meanings, this distinction is irrelevant, and therefore they must be regarded as variants of one and the same sign. Another example: the distinction between signs [nju] and [nu] is not correlated with a distinction between their meanings. Therefore, these signs are variants of the same sign denoted by a sequence of letters <u>new</u>.

The Principle of Semiotic Relevance has empirically testable consequences. The validity of this principle depends on whether counterexamples could be found which would invalidate it. What evidence would serve to invalidate it?

Consider ambiguity. If the Principle of Semiotic Relevance were invalid, then we could distinguish different meanings no matter whether or not they were correlated with different sound sequences. Let us take the sentence <u>I have called the man from New York</u>. This is an ambiguous sentence; it has two very different meanings: 'I have called the man who was from New York;' or 'From New York, I called the man.' One cannot distinguish between the two meanings because they are not correlated with different sequences of sounds. Or take a case of phonological neutralization: German [rat] may

mean either <u>Rad</u> or <u>Rat</u>, but speakers of German do not distinguish between these meanings because they are not correlated with different
sound sequences. This ambiguity can be resolved by different contexts only because there exist oppositions such as <u>Rade:Rate.</u>

No explanation of puns or jokes would be possible without an
understanding of the Principle of Semiotic Relevance. Consider:
<u>Why did the little moron throw the clock out the window?</u> <u>To see</u>
<u>time fly.</u> Or: <u>Why didn't the skeleton jump out of the grave?</u>
<u>Because he didn't have the guts to.</u> The ambiguity of the expres-
sions <u>the time flies</u> and <u>to have guts to do something</u> can be explained by the absence of a correlation between the difference in
meanings and difference between sound sequences.

In all the above examples, a suspension of a distinction between
different meanings is the result of the suspension between a distinction between different sound sequences.

We must distinguish lexical and grammatical meanings. A <u>gramma-</u>
<u>tical</u> <u>meaning</u> is defined as a meaning which is obligatory in a lan-
guage.[3] For example, when we say:

> (1) <u>The boy bought the book</u>

we mean that a definite single boy in the past bought a definite
single book. In conveying our message, we are obliged to assign the
meaning 'definite' or 'indefinite,' 'singular' or plural' to the
nouns, and the meaning 'present,' 'past' or 'future' to the verb; we
have to make a choice between obligatory alternatives. The meanings
'definite' and 'indefinite,' 'singular' and 'plural,' 'present,'
'past' and 'future' are grammatical in English, but they may be non-
grammatical in other languages. Thus, the Russian translation of
the above English sentence:

> (2) <u>Mal'čik kupil knigu</u>

does not involve the obligatory distinction of 'definite' and 'indefinite'; therefore, these meanings are not grammatical in Russian.
If we translate this sentence into some other languages, say Chinese,
we will see that the distinction of 'present,' 'past' and 'future'
may not be obligatory, and so these meanings are not grammatical in
these languages. In some languages 'singular' and 'plural' are not
grammatical either. In keeping with the Principle of Semiotic Relevance, the distinction between grammatical meanings must correlate
with the distinction between different signs: different grammatical
meanings must be expressed by different signs.

Meanings which are not obligatory in a language are called
<u>lexical.</u>

In view of the distinction between grammatical and lexical meanings, the essential difference between languages is not in what may
and may not be expressed, but in what must be expressed. Practi-
cally, any meaning can be expressed in any language. It is crucial
to ask the question: Is a given meaning in a given language obligatory, that is grammatical, or not obligatory, that is lexical?

Grammatical meanings are expressed by <u>grammatical coding devices</u>.
By a grammatical coding device I mean any sign of a grammatical meaning. Two types of grammatical coding devices must be distinguished:
morphological and syntactic ones. Morphological coding devices are:
affixation (e.g. <u>love-er</u> derived from <u>love</u>), modification (e.g.
<u>take</u>►<u>took</u>), transference (also called conversion; e.g. <u>to bite</u>►<u>a
bite</u>), truncation (e.g. in Russian, <u>belyj</u> 'white'►<u>bel</u> 'is white'),
iteration (e.g. <u>riff-raff</u>). Syntactic coding devices are: prepositions and other non-autonomous words, word order, sentence intonation.

One must distinguish between syntactic coding devices and syntax
proper. Consider the following English sentence:

(3) <u>Mary is reading the newspaper to Ann</u>

and its Russian counterpart:

(4) <u>Mari-ja čitaet gazet-u Ann-e</u>

The two sentences have a similar syntactic structure; it consists
of a predicate and three terms: subject, direct object, indirect
object. This structure is represented by different coding devices.
English uses two types of syntactic coding devices. English uses
two types of syntactic coding devices: 1) word order and 2) prepositional markers; the subject is designated by its position before
the predicate, direct object, by its position after the predicate,
and indirect object is designated by the prepositional marker <u>to</u>.
Russian, on the contrary, represents the same syntactic structure by
morphological coding devices: the subject is designated by the nominative case suffix -<u>ja</u>, direct object is designated by the suffix
of the accusative case -<u>u</u>, indirect object is designated by the
suffix of the dative case -<u>e</u>.

In view of the distinction of grammatical meanings, and coding
devices expressing the grammatical meanings, the Correspondence Hypothesis must be restated in a more general form:

> Ergative and accusative coding devices (either morphologi-
> cal or syntactic) correlate with distinct syntactic con-
> structions, and ergative constructions differ from accusa-
> tive constructions both in coding devices and syntax.

In the next section, an empirical test of this hypothesis will be presented.

2.   The Markedness Law, the Dominance Law, and the Law of Duality.

The starting point for the Semiotic Theory of Ergativity is the fact that the notions 'subject' and 'direct object' do not hold when applied to the description of ergative languages. To see this, consider, for instance, Anderson's provocative analysis of the notion of 'subject' in ergative languages (Anderson, 1976).

To investigate the notion 'subject' in ergative languages, Anderson examines the application of rules such as Equi-NP and Subject raising in these languages. Anderson claims that the particular class of NPs in ergative language--the absolutives in intransitive and the ergatives in transitive constructions--to which the rules in question apply constitute exactly the syntactic class denoted by the term 'subject' in the accusative languages. Anderson concludes that the notion ' subject' is the same in most ergative and accusative languages, and, therefore, most ergative languages do not differ in their syntax from accusative languages, the only difference between them being 'a comparatively trivial fact about morphology.' (Anderson, 1976:17).

A close examination of the relevant facts shows that in reality the two classes of NPs in the accusative and ergative languages are very different, and, therefore, the term 'subject' is inappropriate for the class of NPs in the ergative languages. To see this, let us turn to Anderson's analysis. It cannot be denied that in most ergative languages, with respect to the application of Equi and Subject raising, ergatives are similar to transitive subject in accusative languages. But does this similarity justify the generalization that in ergative languages the NPs to which Equi and Subject raising apply belong to the class of subjects?

To answer this question, we must bear in mind that the subject is a cluster concept, that is a concept that is characterized by a set of properties rather than by a single property. The application of Equi and Subject raising is not a sufficient criterion for determining the class of subjects. Among other criteria there is at least one that is crucial for characterizing the class of subjects. I mean the fundamental Criterion of the Non-Omissibility of the Subject. A non-subject may be eliminated from a sentence which will still remain a complete sentence. But this is normally not true of the subject. For instance:

(5)   a.   Peter sells fruit (for a living).
      b.   Peter-sells (for a living).
      c.   *Sells fruit (for a living).

The Criterion of the Non-Omissibility of the Subjects is so important that some linguists consider it a single essential feature for the formal characterization of the subject (Martinet, 1975:219-224). This criterion is high on Keenan's Subject Properties List (Keenan, 1976:313; Keenan uses the term 'indispensability' instead of 'non-omissibility').

Omissibility should not be confused with ellipsis. Ellipsis is a rule of eliminating syntactic units in specific contexts, and the opposition omission:ellipsis is one of the important aspects of the syntactic structure of any natural language.

When we aply a rule of ellipsis, we can always recover the term which was dropped by ellipsis; but an omitted term cannot be recovered: if in John ate clams we omit clams, we get John ate, and clams cannot be recovered because, starting from John ate, we do not know which noun was omitted.

Every languages has specific rules of ellipsis. Thus, in Latin the rule of ellipsis requires use of predicates without personal pronouns. In Latin we say normally Amo, 'I love.' This is a complete sentence with a predicate and a subject which is implied by the context. In Latin we may say Ego amo only in case we want to place stylistic stress on the personal pronoun. In Russian, the rules of ellipsis are directly opposite to the rules of ellipsis in Latin: the normal Russian sentence corresponding to Latin Amo is Ja ljublju without the ellipsis of the personal pronoun; but if we want to place stylistic stress, we use ellipsis and say Ljublju. The Russian Ljubjlju is as complete a sentences as Latin Amo. Both sentences have a subject and a predicate.

Mel'čuk characterizes the Criterion of Non-Omissibility as follows (he uses the term 'nondeletability' as a synonym of 'non-omissibility'):

> Deletability (called also dispensability; see Van Valin, 1970:690) is a powerful and reliable test for the privileged status of any NP: if there is in the language only one type of NP which cannot be omitted from the surface-syntactic structure of the sentence without affecting the grammaticality of the latter or its independence from the linguistic content, then this NP is syntactically privileged. Note that in English it is GS [grammatical subject] (and only GS) that possesses the property of non-deletability among all types of NPs. To put it differently, if a grammatical sentence in English includes only one NP it must be GS. (Imperative sentence like Read this book!, etc. do not contradict the last statement. Based on such cases as Wash yourself/yourselves!, Everybody stand up!, and the like, a GS-you-is postulated in their surface-syntactic structures, where this GS cannot be omitted. It

does not appear in the actual sentence following some rules of ellipsis. As for pseudo-imperative sentences of the type Fuck you, bas-
tard!, these are explained away in the penetrating essay by Quang
(1971).

(Mel'čuk, 1983:235-236)

The Criterion of the Non-Omissibility of the Subjects excludes
the possibility of languages where subjects could be eliminated from
sentences. Yet this is precisely the case with ergative languages,
if we identify ergatives with transitive subjects and absolutives
with intransitive subjects in intransitive constructions and with
transitive objects in transitive constructions. In many ergative
languages we can normally eliminate ergatives, but we cannot eliminate absolutives from transitive constructions. Here is an example
from Tongan (Churchward, 1953:69):

(6)   a.   'Oku taki au 'e Siale.

           'Charlie leads me.'

      b.   'Oku taki au.

           'Leads me (I am led).'

'e Siale in (6a) is an ergative. It is omitted in (6b) which is a
normal way of expressing in Tongan what we express in English by
means of a passive verb (Tongan does not have passive).

Notice that in accusative languages the opposition subject:direct
object is normally correlated with the opposition active voice:pas-
sive voice, while ergative languages normally do not have the opposition active voice:passive voice. This has significant consequences.
In order to compensate for the lack of the passive, ergative languages use the omission of ergatives as a normal syntactic procedure
which corresponds to passivization in accusative languages (an absolutive in a construction with an omitted ergative corresponds to a
subject in an agentless passive construction in an accusative language) or use focus rules which make it possible to impart prominence
to any member of a sentence (in this case either an absolutive or an
ergative may correspond to a subject in an accusative language).
Here is an example of the application of focus rules in Tongan
(Churchward, 1953:67):

(7)   a.   Na'e tamate'i 'e Tevita 'a Kolaiate.

           'David killed Goliath.'

      b.   Na'e tamate'i 'a Kolaiate 'e Tevita.

           'Goliath was killed by David.'

Sentence (7a) corresponds to <u>David killed Goliath</u> in English, while (7b) corresponds to <u>David was killed by David</u>. In the first case, ergative <u>'e Tevita</u> corresponds to the subject <u>David</u> in the active construction, while in the second case, absolutive <u>'a Kolaiate</u> corresponds to the subject <u>Goliath</u> in the passive. The focus rule gives prominence to the noun which immediately follows the verb, that is to <u>'e Tevita</u> in (7a) and to <u>'a Kolaiate</u> in (7b).

It should be noted that focus rules by no means can be identified with passivization rules. They are only a compensation, a substitute for passivization rules.

In Tongan, as in many other ergative languages, we are faced by a serious difficulty resulting from the following contradiction: if the class of subjects is characterized by the application of Equi and Subject Raising, then ergatives are subjects in transitive constructions and absolutives are subjects in intransitive constructions; but if the class of subjects is characterized by the Criterion of the Non-Omissibility of the Subject, then only absolutives can be subjects in transitive constructions.

Since we cannot dispense with either of these criteria, this creates a contradiction in defining the essential properties of the subjects.

One might question whether we could dispense with either criterion. What if we choose to define subject in terms of one and disregard the other? The answer is that no essential criterion can be dispensed with in a theoretically adequate definition because any theoretically adequate definition must include all essential features of the defined concept. We could dispense with one of these criteria only if we considered one of them inessential, that is relating to an accidental feature of subject. But, as is well known, non-omissibility is a non-accidental, permanent, essential feature of subject: subject is non-omissible because it is a syntactically distinguished, central, highly privileged term of a sentence in a given language. On the other hand, the behavior of subject with respect to Equi-NP Deletion and Subject Raising is also essential; therefore, we cannot dispense with this criterion, either. The two criteria are essential in defining the notion of subject, but at the same time they contradict one another when the notion of ergative is equated with the notion of subject. This contradiction I call the <u>paradox of ergativity</u>.

The first step towards solving the paradox of ergativity is to recognize that <u>ergative</u> and <u>absolutive</u> cannot be defined in terms of <u>subject</u> and <u>object</u>, but, rather, that these are distinct primitive syntactic funtions.

The syntactic functions 'absolutive' and 'ergative' should be

strictly distinguished from morphological cases 'absolutive' and
'ergative.' First, some languages, like Abkhaz or Mayan languages,
are case-less, but they have syntactic functions 'absolutive' and
'ergative.' Second, the syntactic function 'ergative' may be
denoted not only by the ergative case but by other oblique cases and
coding devices (including word order). Of course, we must establish operational definitions of syntactic functions 'absolutive'
and 'ergative.' An instance of such an operational definition is
presented in section 6.

Let us now compare the two particular classes of terms to which
the syntactic rules in question apply: 1) the intransitive and tran-
sitve subjects in accusative languages and 2) the absolutives in intransitive clauses and ergatives in transitive clauses in ergative
languages. The first class is homogeneous with respect to non-omissibility (both intransitive and transitive subjects are non-omissible), but the second class is heterogeneous with respect to this
property of terms; the absolutives are non-omissible while the ergatives are omissible terms. The heterogeneity of the class of terms
to which the syntactic rules in question apply in ergative languages
is an anomaly that calls for an explanation. We face the problem:
How to resolve the contradiction that the ergative, which is the
omissible term of a clause, is treated under the rules in question
as if it were the non-omissible term?

In order to solve our problem, let us consider more closely the
syntactic oppositions ergative:absolutive and subject:object.

Both these oppositions can be neutralized. Thus, ergatives and
absolutives contrast only as arguments of two-place predicates. The
point of neutralization is the NP position in a one-place predicate
where only an absolutive occurs.

The question arises: What is the meaning of the syntactic functions 'ergative' and 'absolutive'?

Ergative means 'agent,' which we will symbolize by 'A'. Absolutive, contrasting with ergative, means' patient'--henceforth symbo-
lized by 'P'. Since in the point of neutralization an absolutive
replaces the opposition ergative:absolutive, it can function either
as an ergative or as an absolutive, contrasting with ergative, that
is, semantically, it may mean either 'agent' (the meaning of an ergative) or 'patient' (the meaning of an absolutive contrasting with
an ergative).

The absolutive is a neutral-negative (unmarked) member of the
syntactic opposition ergative:absolutive and the ergative is a positive (marked) member of this opposition. This may be represented by
the following diagram:

(8)

The meaning of subject and object is defined in terms of A and P as
follows. Object means P; subject contrasting with object means A.
In the point of neutralization, subject replaces the opposition
subject:object; therefore, it can function either as subject contrasting with object or as object, that is, semantically, it may mean
either 'agent' (the meaning of subject contrasting with object) or
'patient' (the meaning of object).

The subject is a neutral-negative (unmarked) member of the syntactic opposition subject:object and the object is a positive
(marked) member of this opposition. This may be represented by the
following diagram:

(9)

The diagram shows:

Subj (A)   ←→   Obj (P)

Subj (A/P)

We come up with opposition unmarked term:marked term. On the
basis of this opposition we establish the following correspondence
between cases in ergative and accusative constructions:

(10) | Opposition of markedness | Ergative construction | Accusative construction |
| --- | --- | --- |
| Unmarked term | Absolutive | Subject |
| Marked term | Ergative | Object |

Examples of the neutralization of syntactic opposition in English
(an accusative language):

(11)   a.   John sells automobiles.

b.   John dances well.

c.   Automobiles sell well.

d.   Automobiles are sold.

In (11a), which is a transitive construction, transitive subject John
is an agent, and transitive object automobiles is a patient. In the
intransitive constructions a subject denotes either an agent, in

(11b), or a patient, in (11c) and (11d).

Examples of neutralization of syntactic oppositions in Tongan
(an ergative language):

(12) a. <u>na'e</u> <u>inu</u>   <u>'a</u>   e   <u>kava</u> <u>'e</u>   <u>Sione</u>

          Past drink Abs. the kava Erg. John

          'John drank the kava.'

      b. <u>Na'e</u> <u>inu</u>   <u>'a</u>   <u>Sione</u>

          Past drink Abs. John

          'John drank.'

      c. <u>Na'e</u> <u>lea</u>   <u>'a</u>   <u>Tolu</u>

          Past speak Abs. Tolu

          'Tolu spoke.'

      d. <u>Na'e</u> <u>'uheina</u> <u>'a</u>   e   <u>ngoúe.</u>

          'The garden was rained upon.'

In (12a) the ergative <u>'e Sione</u> denotes an agent and the absolutive
<u>'a e kava</u> denotes a patient. In (12b) the transitive <u>inu</u> is used as
an intransitive verb; therefore, we have here the absolutive <u>'a Sione</u>
instead of the ergative <u>'e Sione</u>. In (12c) the absolutive <u>'a Tolu</u>
denotes an agent. In (12d) the absolutive <u>'a e ngoúe</u> denotes a
patient.

Speaking of the neutralization of syntactic oppositions in ergative languages, we should not confuse ergative languages with <u>active</u>
(agentive) languages. Some linguists consider active languages to
be a variety of ergative languages. This view is incorrect: as a
matter of fact, active languages are polarly opposed both to ergative
and accusative languages. Thus, Sapir distinguished the active construction (typified by Dakota) from both the ergative construction
(typified by Chinook) and the accusative construction (typified by
Paiute) (Sapir, 1917:86). The ergative and accusative constructions
are both based upon a verbal opposition <u>transitive:intransitive,</u>
while for the active construction the basis of verbal classification
is not the opposition <u>transitive:intransitive</u> (which is absent
here), but rather a classification of verbs as <u>active</u> and <u>inactive</u>
(<u>stative</u>). In a series of publications G. A. Klimov has demonstrated the radical distinctness of the active construction from the ergative and accusative constructions and has provided a typology of

the active constructions (Klimov, 1972; 1973; 1974; 1977). The
notion of the active construction as radically distinct from the
ergative construction is shared by a number of contemporary linguists
(see, for example, Aronson, 1977; Dik, 1978; Kibrik, 1979). In
active languages the verbal classification as active and inactive
correlates with the formal opposition of terms active (agent):
inactive (patient). Since this opposition is valid both for two-
place and one-place predicates (a one-place predicate can be combined with a noun in an active or in an inactive case), the NP
position in a one-place predicate cannot be considered a point of
neutralization of the opposition active:inactive. So, the notion of
syntactic neutralization is inapplicable to active constructions.
Therefore, the discussion below of the consequences of the syntactic neutralization in ergative languages does not apply to active
languages, where syntactic neutralization is absent.

The markedness relation between absolutive and ergative, subject
and object, is defined by the Markedness Law:

> Given two semiotic units A and B which are members of a
> binary opposition A:B, if the range of A is wider than the
> range of B, then the set of the relevant features of A if
> narrower than the set of the relevant features of B.

(By the range of a semiotic unit I mean the sum of its syntactic positions. The term 'relevant feature' means here an essential feature which constitutes a part of the definition of a semiotic unit.)

The binary opposition A:B characterized by the Markedness Law is
called the markedness relation between A and B; A is called the unmarked term and B the marked term of this opposition.[4]

Under the Markedness Law, absolutives and subjects are unmarked
terms, and ergatives and objects are marked ones, because absolutives
and subjects occur both with one-place and two-place predicates,
while ergatives and objects normally occur only with two-place predicates. 'Agent' is a relevant feature of ergative and 'patient' is
a relevant feature of object, but neither 'agent' nor 'patient' are
relevant features of absolutive and subject: depending on the context, absolutive and subject may mean either 'agent' or 'patient.'

The relation of markedness in a sentence is characterized by the
Dominance Law:

> The marked term in a sentence cannot exist without the un-
> marked term, but the unmarked term may exist without the
> marked term.

A corollary of the Dominance Law is that the unmarked term of a
sentence is its central, independent term. By contrast, the marked

term of a sentence is its marginal, dependent term. I will call the
unmarked term the <u>primary term</u> and the marked term, the <u>secondary
term</u>. The primary term is the only obligatory term of a sentence.

The Dominance Law explains why the marked term is omissible and
the unmarked term is non-omissible in a clause representing the opposition <u>unmarked term:marked term</u>. The marked term is omissible because the unmarked term does not entail the occurrence of the marked
term. And the unmarked term is non-omissible because the marked term
entails the occurrence of the unmarked term, that is, it cannot occur
without the unmarked term.

It is to be noted that the Dominance Law makes an empirical claim
that must be validated by empirical research. But this law, as any
other linguistic law, is an idealization of linguistic reality in
the same sense as physical laws are idealizations of physical reality. Laws and factual statements have different logical status. Just
as in physics empirical research discovers empirically explicable
deviations from physical laws, so in linguistics empirical research
has to discover empirically explicable deviations from linguistic
laws. Empirically explicable deviations from a law should not be
confused with real counterexamples that undermine the law. Thus, in
(5) and (6) I gave some examples of the omissibility and non-omissibility of terms which can be explained by the Dominance Law. But
it is easy to find apparent counterexamples to this law. For example, one could produce a sentence like <u>He put his books in order</u>
as a counterexample to this law; but here we have a semantically
explicable deviation from the law rather than a counterexample. If
we analyze such sentences, we discover a semantic constraint on the
omissibility of direct object: <u>if the meaning of the transitive
predicate is incomplete without the meaning of the direct object,
the direct object cannot be omitted.</u> This constraint has nothing to
do with the syntactic structure of a sentence; it belongs in the
realm of semantics, which gives an independent characterization of
the notion 'incomplete meaning.' There may be found other apparent
counterexamples to the law which actually are derivations explicable
by the rules of ellipsis or some other clearly defined constraints.
We can now formulate a law which I call the <u>Law of Duality</u>:

> The marked term of an ergative construction corresponds to
> the unmarked term of an accusative construction, and the
> unmarked term of an ergative construction corresponds to
> the marked term of an accusative construction; and vice
> versa, the marked term of an accusative construction cor-
> responds to the unmarked term of an ergative construc-
> tion, and the unmarked term of an accusative construction
> corresponds to the marked term of an ergative construction.

An accusative construction and an ergative construction will be
called <u>duals</u> of each other.

The marked and unmarked terms in accusative and ergative constructions are polar categories, like, for example, positive and
negative electric charges; a correspondence of unmarked terms to
marked terms and of marked terms to unmarked terms may be compared
to what physicists call 'charge conjugation,' a change of all plus
charges to minus and all minus to plus.

The Law of Duality characterizes a fundamental contradiction
between the syntactic function and the meaning of syntactic terms.
Thus, although by its syntactic function absolutive is a primary
term, by its meaning it corresponds to direct object which is syntactically a secondary term. And although by its syntactic function
ergative is a secondary term, by its meaning it corresponds to subject, which is syntactically a primary term.

3.   The Paradox of Ergativity and Functional Superposition

Let us now turn to our main problem which, in the light of the
Markedness Law, may be restated as follows: How to resolve the contradiction that ergative, which is the secondary (marked) term of a
clause, is treated under the above rules as if it were the primary
(marked) term of the clause? This is a contradiction implied by the
paradox of ergativity, formulated on page 177.

In order to resolve the paradox of ergativity, we must introduce
a notion which I call functional superposition. This notion characterizes a syntactic unti which has its own characteristic syntactic
function, but in addition has taken on the function of another syntactic unit, so that this function has been superposed on the characteristic function. For example, the characteristic syntactic function of an adjective is to be a modifier of a term, as in blind
people, but an adjective can also take on the syntactic function of
a term, as in the blind deserve our help. The characteristic function of a noun is to be a term, but a noun can also take on the syntactic function of an adjective, that is the syntactic function of a
term modifier, as in money market.

The important thing to notice is that a syntactic unit that
takes on a new syntactic function does not lose its characteristic
function. The characteristic syntactic function and syntactic function superposed on it constitute a syncretic functional property of
the syntactic unit. Thus, in the blind deserve our help, although
the adjective blind has taken on the function of a term, it remains
an adjective. In money market, the noun money has taken the syntactic function of an adjective, and yet it remains a noun.

In Russian, the characteristic syntactic function of the instrumental case is to be an adverbial, but in addition it can take
on the function of the direct object, as in Ivan upravljaet zavodom
'John manages a factory.' This sentence can be passivized - zavod

upravljaetsja Ivanom, 'The factory is managed by John'--because the
instrumental in the active construction functions as an accusative
case. The characteristic function of the accusative case is to be
the direct object, but in addition it can take on the function of an
adverbial, as in On rabotal celyj den' 'He worked all day long.' In
this sentence, the accusative celyj den' may be replaced by an
adverbial of time, such as utrom 'in the morning,' večerom 'in the
evening,' etc. Compare: On rabotal večerom 'He worked in the
evening.' This syntactic behavior of the accusative shows that it
functions as an adverbial.

The characteristic function of Russian dative is the role of an
indirect object, but there is a large class of predicates which
superpose the function of the subject on it. For example:

(13)   Vidja čto proisxodit, emu stydno   za   svoego   brata

       seeing what happens him-Dat is-ashamed for his-Refl
                                                          brother
       'Seeing what is happening, he is ashamed for his brother.'

In (13) the dative emu has its characterisitc function of the indirect object, but in addition, the predicate stydno superposes on it
the function of the subject, so that emu has three properties of the
subject: 1) it controls Equi in a participle construction vidja čto
proisxodit, 2) it serves as an antecedent of the reflexive svoego,
3) it precedes the predicate stydno (although Russian has a free
word order, subjects normally precede predicates, while direct and
indirect and indirect objects follow them).

The notion of functional superposition throws light on our problem. The important thing to notice is that only primary terms may
appear in the intransitive clauses. An identification of a term of
the transitive clause with the primary term of the intransitive
clause involves a superposition of the function of the primary term
in the intransitive clause on the function of the given term of the
transitive clause. Three possibilities are open: 1) only the primary term of a transitive clause may be identified with the primary
term of an intransitive clause (no superposition); 2) only the
secondary term of a transitive clause may be identified with the
primary term of an intransitive clause (a superposition of the function of the primary term); 3) both the primary and the secondary
terms of a transitive sentence may be identified with the primary
term of an intransitive sentence (no superposition or the superposition of the function of the primary term of an intransitive clause
on the secondary term of the transitive clause).

Accusative languages realize only the first possibility: both
intransitive subject and transitive subject are primary terms. But
all three possibilities are realized in ergative languages: 1) the

syntactic rules in question are stated with reference to only absolutives in intransitive and transitive clauses (Dyirbal); 2) the syntactic rules in question are stated with reference to absolutives in
intransitive clauses and ergatives in transitive clauses (Basque);
3) the syntactic rules in question are stated with reference to absolutives in intransitive clauses and to absolutives and ergatives in
transitive clauses (Archi, a Daghestan language; Kibrik, 1979:71-72).

The notion of functional superposition in ergative constructions
should not be confused with the notion of the pivot introduced by
Dixon (1979). These notions have nothing in common. Rather, they
are opposed to each other. While the notion of functional superposition, characterizing a syncretic property of syntactic units, reveals
the radical distinctness of the syntactic structure of the ergative
construction from the syntactic structure of the accusative construction, 'pivot' is nothing but a descriptive term which glosses over
this distinctness. Dixon uses symbol S to denote intransitive subject, symbol A to denote transitive subject and symbol O to denote
transitive object. Syntactic rules may treat S and A in the same
way, or they may treat S and O in the same way; "we refer to S/A
and S/O pivots respectively (Dixon, 1979:132)." Dixon writes:
"Many languages which have an ergative morphology do not have ergative syntax; instead syntactic rules seem to operate on an 'accusative' principle treating S and A in the same way (Dixon, 1979:63."
Referring to Anderson (1976), Dixon relegates Basque and most other
ergative languages (except Dyirbal and a few others) to a class of
languages which have ergative morphology, but accusative syntax.
Thus, in spite of a different terminology, Dixon shares the same
view with Anderson and other linguists who claim that the syntactic
structure of ergative constructions in most ergative languages is
identical with the syntactic structure of accusative constructions.

Why does functional superposition occur in ergative constructions and does not occur in accusative constructions? This fact
may be explained by a semantic hypothesis advanced on independent
grounds. According to this hypothesis based on the semiotic principle of iconicity, the sequence agent-patient is more natural than
the sequence patient-agent, because the first sequence is an image
of a natural hierarchy according to which the agent is the starting
point and the patient is the end point of an action. The semantic
hierarchy agent-patient coincides with the syntactic hierarchy transitive subject-direct object in accusative languages because transitive subject denotes agent and direct object denotes patient. But
this semantic hierarchy contradicts the syntactic hierarchy absolu-
tive-ergative because absolutive, being syntactically a primary term
denotes 'patient', which is semantically a secondary term, and ergative being syntactically a secondary term denotes 'agent,' which is
semantically a primary term. Hence, under the pressure of the semantic hierarchy, agent-patient functional superposition assigns to the
ergative the role of a syntactically primary term.

The above hypothesis seems to be plausible. But I want to stress that whether or not one accepts this hypothesis, the syntactic fact of functional superposition is established on the basis of the syntactic behavior of syntactic terms independently of any semantic hypotheses. We must strictly distinguish two types of explanation: 1) syntactic explanation and 2) semantic explanation. Using the notion of functional superposition in conjunction with the Markedness Law, the Dominance Law, and the Law of Duality, we explain the syntactic behavior of the terms in ergative constructions and discover a fundamental difference between the syntactic structures of the ergative and the accusative. This is the syntactic part of our research. On the basis of the syntactic research we may pose interesting questions which call for semantic or pragmatic explanation, but we should always bear in mind that syntactic phenomena must be explained by syntactic laws in the first place.

I propose the following definition of the notions 'accusative construction' and 'ergative construction':

The accusative construction and ergative construction are two representations of the abstract transitive/intransitive clause pattern: primary term+transitive predicate+secondary term / primary term+intransitive predicate.

2.   Primary term is represented by subject in the accusative construction and by absolutive in the ergative construction. Secondary term is represented by direct object in the accusative construction and by ergative in the ergative construction.

3.   The abstract clause pattern and its representations are characterized by the Markedness Law and the Dominance Law.

4.   There is a correlation between the accusative and ergative constructions characterized by the Law of Duality.

5.   In the ergative construction the primary and secondary terms may exchange their functions, so that the function of the primary term is superposed on the secondary term and the function of the secondary term is superposed on the primary term.

In order to make the rules Equi and Subject Raising valid both for accusative and ergative languages, we have to replace them by more abstract rules: Equi-Primary Term Deletion and Primary Term Raising. The generalizations expressed by these abstract rules solve the problem raised by Anderson. Contrary to his claim, there are neither subjects, nor direct objects in ergative languages: ergative and absolutive are distinct primitive syntactic functions. What ergative and accusative constructions have in common is that

they are different realizations of the abstract construction primary
term:secondary term. The new abstract rules represent a correct
generalization which cannot be captured in terms of subject and
direct object.

The rules Equi-Primary Term Deletion and Primary Term Raising lay
bare a parallelism between the syntactic structure of the ergative
construction in Dyirbal and the syntactic structure of the accusative
construction but a sharp difference between the syntactic structure
of the ergative construction in Basque and similar ergative languages
and the accusative construction. Since in Dyirbal the rules in
question apply only to absolutives both in the intransitive and
transitive constructions, the only difference between the ergative
construction in Dyirbal and the accusative construction, say in
English, boils down to polar semantic interpretations: while primary terms in Dyirbal ergative constructions denote patients and
secondary terms, agents, in any accusative construction, quite the
reverse, primary term denotes agent and secondary term, patient.

Consider an example of an ergative construction from Dyirbal
(Dixon, 1972):

(14) yabu   ŋuma + ŋgu   buṛa + n

      mother   father-Agent   see-Past
      primary,   secondary,   predicate
      patient   agent

'Mother was seen by father.'

Compare (14) with the English sentence:

(15) Father   saw mother

      primary,   secondary,
      agent   patient

We see the syntactic function of yabu denoting a patient in (14) is a
counterpart of the syntactic function of father denoting an agent in
(15), while the syntactic function of ŋuma + ŋgu denoting an agent in
(14) is a counterpart of the syntactic function of mother denoting a
patient in (15).

Now, if we turn to a comparison of Basque and English, we discover a sharp difference between the syntactic structure of the accusative and the ergative constructions. Consider examples from
Basque discussed in Anderson (1976:12).

The deleted term from the embedded clause in the underlying structure
had to be the absolutive in (16) and ergative in (17).

(16)   <u>dantzatzerat</u>   <u>joan da</u>

         dance-infn-<u>to</u> go he-is

         'He has gone to dance.'

(17)   <u>txakurraren</u> <u>hiltzera</u>   <u>joan nintzen</u>

         dog-def-gen kill-infn-<u>to</u> go I-was

         'I went to kill the dog.'

In (17) the syntactic structure of the embedded ergative clause
is very different from the syntactic structure of the embedded accusative clause in English because the deleted ergative in the former
has a syntactic behavior which is sharply distinct from the syntactic
behavior of the deleted subject in the latter. While subject is an
intrinsically primary term, ergative is an intrinsically secondary
term. In (17) ergative functions as a primary term, but the function
of the primary term is superposed on the ergative whose characteristic intrinsic syntactic function is to serve as a secondary term.
Since in (17) ergative functions as a primary term, the function of
the secondary term is superposed on absolutive whose characteristic
function is to serve as a primary term. The important thing to
notice is that in transitive clauses similar to (17) neither ergative nor absolutive lose their characteristic function, and, consequently, absolutive can never be deleted in these clauses, while
ergative can, in accordance with the Dominance Law, which holds that
secondary terms presuppose primary terms, while primary terms do not
presuppose secondary terms.

To sum up, there is a parallelism and at the same time a sharp
difference between the syntactic structures of the ergative and accusative constructions. In order to do justice to the parallelism
and the difference, we have to state our generalizations and rules
in terms of the abstract notion 'primary term' and 'secondary term.'
The rules Equi-Primary Deletion and Primary Term Raising capture
what ergative and accusative constructions have in common and at the
same time reflect the laws and principles characterizing a profound
difference between the two types of syntactic constructions.

Thus, we have come to a conclusion which is diametrically opposite to Anderson's view on ergativity. Contrary to his claim that
the majority of ergative languages, with the exception of Dyirbal
and a few similar languages, have the same syntactic structure as
accusative languages, we claim that the syntactic structure of the
majority of ergative languages sharply differs from the syntactic
structure of accusative languages, with the exception of Dyirbal,
whose syntactic structure exhibits a parallelism with the syntactic
structure of accusative languages.

4. Explication of the Concepts of the Semiotic Theory of
    Ergativity through Their Reduction to the Primitive
    Concepts of Applicative Universal Grammar

I defined the notions 'ergative construction' and 'accusative
construction' in terms of eight notions: one-place predicate, two-
place predicate, primary term, secondary term, subject, direct
object, absolutive, and ergative. It was shown that subject and
direct object , absolutive and ergative are non-universal notions:
rather, they are different interpretations of abstract notions 'primary term' and 'secondary term,' which are truly universal.

Can some of these eight notions be further reduced to simpler
notions?

We can define the notions 'primary term,' secondary term,' intransitive predicate,' 'transitive predicate' by reducing them to
three primitive syntactic notions: 'sentence,' 'term,' and 'operator.'

It can be done if we turn to an abstract system provided by
Applicative Grammar (henceforth AG).

AG is a linguistic theory which has integrated two hitherto
separate and independent formal system: 1) the formal system of
categorial grammars and 2) the formal system of the theory of com-
binators in mathematical logic (Curry and Feys, 1958). The integrated formal system I call the genotype language.[5]

I have put forward the notion of the genotype language as a hypothetical simple sign system which underlies natural languages and
controls their functioning. The genotype language is the common
semiotic basis of all natural languages. It is an essential step in
the quest for the ultimate invariants of language.

Here are the main ideas of AG.

Any speech act involves communication and three basic items connected with it: the sender, the receiver and the external situation.
According to whether the communication is oriented towards one of
these three items, it can have, respectively, an expressive, vocative, or representational function.

I leave the expressive and vocative functions for the present and
focus on the representational function of the communication. If we
abstract from everything in the language used which is irrelevant to
the representational function of the communication, we have to recognize as essential only three classes of linguistic expressions:
1) the names of objects, 2) the names of situations, c) the means of
constructing the names of objects and the names of situations.

Names of objects are called <u>terms</u>. For example, the following expressions are used as terms in English: <u>a car</u>, <u>a gray car</u>, <u>a small gray car</u>, <u>a small gray car he bought yesterday</u>. Terms should not be confused with nouns. Noun is a morphological concept whereas term is a functional concept. Languages without word classes lack nouns but still have terms.

<u>Sentences</u> serve as the name of situations.

The means for constructing the names of objects and the names of situations are the expressions called <u>operators</u>. An operator is any kind of linguistic device which acts on one or more expressions called its <u>operands</u> to form an expression called its <u>resultant</u>. For example, in the English expression:

(18)   The <u>hunter killed</u> the <u>bear</u>.

the word <u>killed</u> is an operator that acts on its operands <u>the hunter</u> and <u>the bear</u>; in <u>gray car</u> the expression <u>gray</u> is an operator that acts on its operand <u>car</u>. If an operator has one operand, it is called a <u>one-place</u> operator; if an operator has n operands, it is called an <u>n-place</u> operator.

In ordinary logic an <u>n</u>-place operator combines with its operands in one step. Ordinary logic treats operands as if they had equally close connection with their operator. But in natural language an operator is more closely connected with one operand than another. For example, a transitive verb is more closely connected with the secondary term than with the primary term. Thus, in the above example (18) the transitive predicate <u>killed</u> is more closely connected with <u>the bear</u> than with <u>the hunter</u>:

(19)   ((<u>killed</u> the <u>bear</u>) the <u>hunter</u>)

Why are <u>killed</u> and <u>the bear</u> more closely connected than <u>killed</u> and <u>the hunter</u>? Because a combination of a transitive predicate with a direct object is equivalent to an intransitive predicate. This is why in some languages this combination can be replaced by an intransitive predicate. For example, in Russian <u>lovit' rybu</u> 'to catch fish' may be replaced by the intransitive verb <u>rybačit'</u> with the same meaning. And, vice versa, an intransitive verb may be replaced by a transitive verb with a direct object. For example, <u>to dine</u> may be replaced by <u>to eat dinner</u>. There is also other evidence for a close connection between a transitive verb and its direct object. Thus, nouns derived from intransitive verbs are oriented toward the subjects of the action (<u>genitivus subiectivus</u>), while nouns derived from transitive verbs tend to be oriented toward the object of the action (<u>genitivus obiectivus</u>). Compare: <u>the dogs bark</u>:<u>the barking of the dogs</u> versus <u>they abducted the woman</u>:<u>the abduction of the woman</u>. The ambiguity of expressions like <u>the shooting of the hun-</u>

ters must be explained by the fact that, although the verb to shoot
is transitive, it can be also used an intransitive verb; we can say
the hunters shoot without specifying the object. Compare: the boy
reads the book:the boy reads. The orientation of nouns derived from
transitive verbs toward the object of the action is a universal tendency observed in typologically very different language groups.

To do justice to the above phenomenon, we must redefine the combination of an n-place operator with its operands as a series of
binary operations: an n-place operator is applied to its first oper-
and, then the resultant to the second operand, and so on. According
to the new definition, an n-place operator combines with its operands in n steps rather than in one step as in ordinary logic. For
example, any transitive predicate, which is a two-place operator,
must be applied to the secondary term, then the resultant to the
primary term. Thus, in the expression (18) transitive predicate
killed must be applied first to the bear, then to the hunter:

(19)   ((killed the bear) the hunter)

This binary operation is called application.

The above informal explanation of the notion of application must
now be presented as a formal statement called the Applicative Principle:

An n-place operator is always represented as a one-place
operator that yields an (n-1)-place operator as its resultant.[6]

Examples of representing an n-place operator as a one-place operator.
Two-place operator killed is represented as a one-place operator
which is applied to its operand the bear and yields that resultant
(killed the bear). The resultant is a (2-1)-place operator, that is
a one-place operator which is applied to another term, the hunter,
and yields the resultant ((killed the bear) the hunter). The new
resultant is a (1-1)-place operator, that is a zero-place operator,
which is a sentence. This sentence is an abstract representation of
the sentence The hunter killed the bear.

Three-place operator gave is represented as one-place operator
which is applied to its operand Mary and yields the resultant (gave
Mary). The resultant is a (3-1)place operator, that is a two-place
operator. This two-place operator is represented, in its turn, as a
one-place operator which is applied to another term, money, and
yields the resultant ((gave Mary) money), which is a (2-1)-place
operator, that is a one-place operator. The latter is applied to
the term John and yields the resultant (1-1)-place operator (((gave
Mary) money) John), which is an abstract representation of the
sentence John gave Mary money.

On the basis of the Applicative Principle I define the formal concepts 'one-place predicate,' 'two-place predicate,' 'three-place predicate' and the formal concepts 'primary term,' secondary term,' 'tertiary term':

Definition 1.   if X is an operator which acts on a term Y to form a sentence Z, then X is a <u>one-place predicate</u> and Y is a <u>primary term</u>.

Definition 2.   If X is an operator which acts on a term Y to form an intransitive predicate Z, then X is a <u>two-place predicate</u> and Y is a <u>secondary term</u>.

Definition 3.   If X is an operator which acts on a term Y to form a transitive predicate Z, then X is a <u>three-place predicate</u> and Y is a <u>tertiary term</u>.

All these definitions can be given in a formal notation.   To do this, I will denote the two primitive notions--sentences and terms-- by symbols <u>s</u> and <u>t</u>, respectively, and will introduce the symbol <u>O</u> to denote the general notion of operator.   Then the rule defining the types of operators is stated:

(20)   If x and y are sentences, terms or operators, then <u>Oxy</u> is an operator.[7]

The symbolic expression <u>Oxy</u> should read: an operator which acts on expression x to form an expression y.

Applying this rule to the primitive notions <u>s</u> and <u>t</u>, we get symbolic expressions defining various types of operators.   In particular:

Symbolic expression <u>Ots</u> denotes one-place predicates.

Symbolic expression <u>OtOts</u> denotes two-place predicates.

Symbolic expression <u>OtOtOts</u> denotes three-place predicates.

In these symbolic expressions the first from the end <u>t</u> denotes a primary term, the second from the end <u>t</u> denotes a secondary term, and the third from the end <u>t</u> denotes a tertiary term.

The primary, secondary and tertiary terms are central terms of a sentence because they are part of its predicative structure.   By contrast, oblique terms are marginal terms of a sentence because they are outside of its predicative structure.   The opposition of a primary and a secondary term constitutes the <u>nucleus</u> of a sentence.   These terms I call <u>nuclear</u>.   The tertiary term is outside of the nucleus

because it is not a member of the markedness relation which is a
basic relation between terms in a sentence.

On the basis of the properties of the terms of a sentence defined
in the framework of AG, we can establish the following hierarchy of
terms which I call the <u>Applicative Hierarchy</u>:

(21)   Primary Term $\gg$ Secondary Term $\gg$ Tertiary Term $\gg$ Oblique Term...

To sum up, at the beginning of this section we asked whether the
syntactic notions 'primary term,' 'secondary term,' 'one-place predicate,' 'two-place predicate' could be reduced to simpler notions.
By applying AG to the analysis of these notions we have reduced them
to three primitive notions: 'sentence,' 'term,' and 'operator.'
Actually, any basic syntactic notion can be reduced to these three
primitive notions (see Shaumyan, 1977; 1985).

The reduction of theoretical concepts to primitive theoretical
concepts is one of the most important tasks of linguistic theory because it leads toward the unification and simplification of its premises. A similar task is faced by any theoretical science. Einstein
once said:

New theories are first of all necessary when we encounter new
facts which cannot be 'explained' by existing theories. But
this motivation for setting up new theories is, so to speak,
trivial, imposed from without. There is another more subtle
motive of no less importance. This is striving toward unification and simplification of the premises of the theory as a
whole.

(Einstein, 1973:332-333)

5.   The Semiotic Theory of Ergativity and the Accessibility Hierarchy

In preceding section I tried to present motivations for the Semiotic Theory of Ergativity. In order to produce additional evidence
in support of this theory, I will examine the Keenan-Comrie Accessibility Hierarchy (Keenan, Comrie, 1977).

In an important study of relative clause formation strategy,
Edward L. Keenan and Bernard Comrie established the Accessibility
Hierarchy which characterizes the relative accessibility ro relative
clause formation of various members of a sentence. In terms of the
Accessibility Hierarchy they state universal constraints on relative
clause formation. According to the Accessibility Hierarchy, processes of relative clause formation are sensitive to the following hierarchy of grammatical relations:

(22)   Subject ⪢ Direct object ⪢ Indirect object ⪢ Oblique NP ⪢

⪢ Possessor ⪢ Object of comparison

where ⪢ means 'more accessible than.'

The positions of the Accessibility Hierarchy are to be understood as specifying a set of possible relativizations that a language may make: relativizations that apply at some point of the
hierarchy must apply at any higher point. The Accessibility Hierarchy predicts, for instance, that there is no language which can
relativize direct objects and not subjects or that can relativize
possessors and subjects, but not direct objects and oblique NPs.

In terms of the Accessibility Hierarchy Keenan and Comrie state
the following universal constraints on relative clause formation:

The Hierarchy Constraints

1.   A language must be able to relativize subjects.

2.   Any relative clause forming strategy must apply
     to a continuous segment of the Accessibility
     Hierarchy.

3.   Strategies that apply at one point of the Accessi-
     bility Hierarchy may in principle cease to apply
     at any lower point.

Constraint (1) states that the grammar of any language must allow
relativization on subjects. For instance, no language can relativize
only locatives or direct objects. Constraint (2) says that a language is free to treat the adjacent positions as the same, but it
cannot skip positions. For example, if a given strategy can apply
to both subjects and locatives, it must also apply to direct objects
and indirect objects. Constraint (3) says that each point of the
Accessibility Hierarchy can be a cut-off point for any strategy that
applies to a higher point.

Here are some examples of the data that support the Hierarchy
Constraints (Keenan, Comrie, 1977:69-75).

Subjects only. Many Malayo-Polynesian languages (for example,
Malagasy, Javanese, Iban, Toba-Batak) allow relativization only on
subjects. Looking at Malagasy, the major relative clause formation
process is basically this: the head NP is placed to the left, followed optionally by an invariable relativizer izay, followed by the
restricting clause. Notice that to relativize a direct object, the
sentence is first passivized so that direct object becomes a subject:

(23) a.   Nahita ny vehivavy ny mpianatra.

           saw   the woman   the student

           'The student saw the woman.'

      b.   ny mpianatra izay nahita ny vehivavy

           the student that saw   the woman

           'the student that saw the woman.'

      c.   *ny vehivavy izay nahita ny mipianatra

           the woman   that saw   the student

           'the woman that the student saw'

      d.   Nohitan'ny   mpianatra ny vehivavy.

           seen-PASS the student the woman

           'The woman was seen by the student.'

      e.   ny vehivavy izay nohitan'ny mpianatra

           the woman   that seen   the student

           'the woman that was seen by the student.'

Subject Direct Object. Some languages, for example, Welsh or
Finnish, allow relativization only on subject and direct objects.
Finnish places the relative clause prenominally, uses no relativiza-
tion marker, and puts the subordinate verb in a non-finite form.
Here is an example from Finnish:

(24) a.   Pöydällä tanssinut   poika oli sairas.

           on-table having danced boy   was sick.

           'The boy who had danced on the table was sick.'

      b.   Näkemäni   poika tanssi pöydällä.

           I-having-seen boy   danced on-table.

           'The boy that I saw danced on the table.'

The Accessibility Hierarchy excludes the possibility of languages
where subjects are less accessible to relativization than objects.

Yet precisely this is the case with Mayan languages if the notion
'ergative construction' is defined on the basis of subject, as is
done by the authors of the Accessibility Hierarchy whose stance is
representative of the views on ergativity. To see this, let us turn
to Comrie's definition of the notion 'accusative construction' and
'ergative construction' (Comrie, 1978:343-350; 1979:221-223).

In speaking about the arguments of one-place and two-place predicates, Comrie uses the symbol S to refer to the argument of a one-
place predicate and symbols A (typically agent) and P (typically
patient) to refer to the arguments of a two-place predicate. Where
a predicate has the argument P, it is called a transitive predicate.
All other predicates, whether one-place, two-place or more than two-
place, are called intransitive. An intransitive predicate can and
usually does have an S, but cannot have an A. Using three primitives
S, A and P, Comrie characterizes syntactic ergativity and syntactic
accusativity (nominativity) as follows:

> In treating ergativity from a syntactic viewpoint, we are look-
> ing for syntactic phenomena in languages which treat S and P
> alike, and differently from A. Syntactic nominativity likewise
> means syntactic phenomena where S and A are treated alike, and
> differently from P. This distinction is connected with the gene-
> ral problem of subject identification: if in a language S and A
> are regularly identified, that is, if the language is consis-
> tently or overwhelmingly nominative-accusative, then we are jus-
> tified in using the term subject to group together S and A; if
> in a language S and A are regularly identified (consistent or
> overwhelming ergative-absolutive system), then we would be justi-
> fied in using the term subject rather to refer to S and P, that
> is, in particular, to refer to P, rather than A, of the transi-
> tive construction.

(Comrie, 1978:343)

In accordance with this characterization, Comrie arrives at the
same conclusion as Anderson: he considers morphologically ergative
languages, such as Basque or Tongan, to be syntactically accusative
because these languages treat S and A alike and differently from P;
he considers Dyirbal syntactically to be ergative because this language treats S and P alike and differently from A.

The weakness of such characterization is that the key notion 'to
treat syntactically alike' is not analyzed adequately. What does it
mean to say that Basque treats S and A alike? If it means only the
application of the rules Equi-NP Deletion and Subject Raising, then,
yes, Basque treats S and A alike, and, therefore, must be regarded,
according to this criterion, as syntactically accusative language.
But there is more to the syntax of the ergative construction than
these rules. If we consider the markedness opposition and syntactic

laws and phenomena associated with this key relation, then we con-
clude that Basque does not treat S an A alike. As a matter of fact,
Comrie's characterization of ergativity runs into the same difficul-
ties as Anderson's claim which we discussed in the previous section.
But let us put aside these difficulties for now and turn to the
Accessibility Hierarchy. Our question is: Can the Accessibility
Hierarchy be regarded as a universal law? In order to answer this
question, let us consider the facts of Mayan languages.

As was conclusively demonstrated by Larsen and Norman (1979), in
Mayan languages syntactic rules apply to ergatives in much the same
way as they do in such languages as Basque and Tongan. Since Mayan
ergatives meet Comrie's criteria of subjecthood, they must be considered subjects, and Mayan languages must be regarded as morphologically ergative, but syntactically accusative languages. Granted
that Mayan ergatives must be defined as subjects, the Accessibility
Hierarchy predicts that if Mayan languages allow relativization on
absolutives, they must allow it also on ergatives. But, contrary to
this prediction, in the Mayan languages of the Kanjobalan, Mamean
and Quichean sub-groups, ergative NPs may not as a rule be relativized (nor questioned or focused), while absolutive NPs can. In
order for an ergative NP to undergo relativization, it must be converted into derived absolutive and the verb intransitivized through
the addition of a special intransitivizing suffix. Here is an example of this process in Aguacatec (Larsen and Norman, 1979:358):

(25)   a. ja   0-0-b'iy   yaaj xna7n

           asp. 3sB-3sA-HIT MAN WOMAN

           'the man hit the woman'

        b. na7 m-0-b'iy-oon   xna7n

           WHO dep. asp.-3sB-HIT-suffix WOMAN

           'Who hit the woman?'

        c. ja   0-w-il   yaaj ye m-0-b'iy-oon   xna7n

           asp. 3sB-1sA-SEE MAN THE dep.asp.-3sB-HIT-suffix WOMAN

        d. yaaj m-0-b'iy-oon   xna7n

           MAN dep.asp.-3sB-HIT-suffix WOMAN

           'it was the man who hit the woman.'

Here -oon is the intransitivizing suffix used to circumvent the
constraints on the extraction of ergatives (the term extraction rules

is a cover term for relativization rules, focus rules, WH-Question).

The features of Mayan languages under discussion closely conform
to those of Dyirbal language, but while Dyirbal absolutive meets
Comrie's criteria of subjecthood, the Mayan absolutive does not.

Dyirbal does not allow relativization on ergatives; instead, the
verb of the relative clause is intransitivized by adding the suffix
ŋa-y , and the ergative is replaced by the absolutive case (Dixon,
1972:100). For instance consider the Dyirbal sentence:

    (26)   <u>yabu</u>   <u>ŋuma+ŋgu</u>   <u>buṛa+n</u>

            mother-ABS father+ERG see+PAST

            'Father saw mother.'

In sentence (26) ergative is marked by -ŋgu. In order to be embedded
into another sentence as a relative clause, sentence (26) must be
antipassivized and ergative ŋuma+ŋgu replaced by absolutive ŋuma+∅.
We may get, for example, the sentence:

    (27)   <u>ŋuma+∅</u>   [buṛal+ŋa+ŋu+∅   yabu+gu]   <u>duŋgara+nʸu</u>

            father+ABS   see-ANTIPASS+REL+ABS mother+DAT cry+PAST

            'Father, who saw mother, was crying.'

We see that the facts of Mayan languages present strong evidence
against the Accessibility Hierarchy. Does it mean that the Accessibility Hierarchy must be abandoned as a universal law? I do not
think so. The trouble with the Accessibility Hierarchy is that it
is formulated as a universal law in non-universal terms, such as subject, direct object, etc. To solve this difficulty, it is necessary
to abandon non-universal concepts, such as subject and direct object,
and replace them by really universal concepts. The key to the solu-
ion of this difficulty is provided by AG.

From the point of view of AG, the Accessibility Heirarchy is a
particular instance of the Applicative Hierarchy established on independent grounds:

    (28)   Primary Term⟹Secondary Term⟹Tertiary Term⟹Oblique Term...

The Applicative Hierarchy is interpreted in accusative languages as:

    (29)   Subject⟹Direct Object⟹Indirect Object⟹Oblique Term...

and in ergative languages as:

(30)   Absolutive ⇒Ergative ⇒Indirect Object⇒Oblique Term...

We see that the confusion of ergatives with subjects in inconsistent with the Acccessibility Hierarchy; which creates an unresolvable difficulty. The treatment of ergatives and subjects as different syntactic functions on the other hand, leads to a deeper understanding of the Accessibility Hierarchy, which results in its restatement on the abstract level in keeping with true basic syntactic universals: primary, secondary and tertiary terms.

The revised Accessibility Hierarchy accounts both for the facts which motivated the original Accessibility Hierarchy and the facts which have been shown to contravene it. The revised Accessibility Hierarchy excludes the possibility of languages where primary terms are less accessible to relativization than secondary terms. And this requirement applies both to accusative languages, where primary terms are interpreted as subjects and secondary terms as direct objects, and to ergative languages, where primary terms are interpreted as absolutives and secondary terms as ergatives. All the facts which support the original Accessibility Hierarchy support also the revised Accessibility Hierarchy. But, besides, the revised Accessibility Hierarchy is supported by the facts, like the above examples from Aguacatec, which contravene the original Accessibility Hierarchy. This is a significant result which shows the importance of the abstract concepts of AG.

In conclusion, I want to dispel a possible misunderstanding of the concepts I have introduced. It was said above that in order to save the Accessibility Hierarchy, it is necessary to abandon the non-universal concepts 'subject' and 'direct object' and replace them by the universal concepts 'primary term' and 'secondary term.' The important thing to notice is that I suggest to replace one set of concepts by another set of concepts rather than one set of terms by another set of terms. The new terms 'primary term' and 'secondary term' designate a very different set of concepts than the concepts designated by the terms 'subject' and 'direct object.' One might argue that we could save the Accessibility Hierarchy by equating subject with absolutive and object with ergative. But this suggestion would obscure the essential difference between the three sets of concepts: 1) primary term:secondary term, 2) subject:direct object, 3) absolutive:ergative.

No matter which terminology we use, we must distinguish these three very different sets of concepts. The second and the third set of concepts are different interpretations (in accusative and ergative languages) of the first truly universal set of syntactic concepts.

6.   Implications of the Correspondence Hypothesis for
      Linguistic Typology

6.1   Ergativity as a Grammatical Category

One fundamental consequence of the Correspondence Hypothesis is that only those ergative processes can he considered formal ergative processes which correlate with ergative morphology.

I propose a broad definition of morphology which includes any coding devices of a language. Under this definition, word order is part of morphology.

Ergative processes may be found in languages which do not have ergative morphology, that is, they are not distinguished by special coding devices. For example, as far as nominalizations are concerned, Russian, an accusative language, has an ergative process: genitive functions as an absolutive and instrumental functions as an ergative (Comrie, 1978:375-376). In French and Turkish, both accusative languages, there are causative constructions which are formed on ergative principles (Comrie, 1976:262-263); in French there are antipassive constructions (Postal, 1977); ergative patterns in various accusative languages are presented in Moravcsik (1978).

Can ergative processes not distinguished by coding devices be considered distinct formal processes?

A language is a sign system. But in a sign system signata cannot exist without signantia, that is without distinct coding devices. True, any natural language is very complex sign system in which there is no one-one correspondence between signantia and signata. Rather, one signans may correspond to many signata, and, vice versa, one signatum may correspond to many signantia. But different signata must belong in the same class if they are not distinguished from one another by at least one distinct coding rule. I propose the concept of the <u>grammatical category</u> which I define as follows:

A <u>grammatical category</u> is a class of grammatical signata that is distinguished from other classes of grammatical signata by at least one distinct coding rule.

In studying natural languages one may discover various linguistic relations. But, if given linguistic relations are not distinguished from one another by at least one distinct coding rule, they belong in the same grammatical category.

Under the proposed definition of the grammatical category, ergativity can constitute a distinct grammatical category in a given language only if it is distinguished from other relations by at least one distinct coding rule.

In order to make my case concrete, I will consider ergativity in Russian. It is claimed that "as far as nominalizations are con-

cerned, Russian has in effect an ergative system (Comrie, 1978:376)."
This claim is based on the following data.

In Russian, passive constructions can be nominalized. For example, we may have:

(31)   a.   <u>Gorod razrušen   vragom</u>
              city has-been-destroyed enemy-by

              'The city has been destroyed by the enemy.'

        b.   <u>razrušenie goroda   vragom</u>

              destruction city-of enemy-by

              'the city's destruction by the enemy.'

(31b) is a nominalization of 31a). In (31b) the genitive <u>goroda</u>
denotes a patient and instrumental <u>vragom</u> denotes an agent, and the
verbal noun <u>razrušenie</u> corresponds to a transitive predicate. This
nominal construction correlates with a nominal construction in which
a verbal noun corresponds to an intransitive predicate and a genitive
denotes an agent, for example:

(32)   <u>priezd vraga</u>
        arrival enemy-of
        'the enemy's arrival'

If we compare (31b) with (32), we can see that the patient in
(31b) and the agent in (32) stand in the genitive (functioning as
an absolutive) while the agent in (31b) stands in the instrumental
(functioning as ergative). Therefore, we may conclude that in
Russian, nominalizations involve ergativity.

Does ergativity constitute a distinct grammatical category in
Russian nominal constructions?

Consider the following example of nominalization in Russian:

(33)   a.   <u>Ivan prenebregaet zanjatijami.</u>

              John neglects   studies

              'John neglects (his) studies.'

        b.   <u>prenebreženie Ivana   zanjatijami</u>

              neglect   John-gen studies-instr

              'John's neglect of studies.'

The surface structure of (33b) is the same as the surface structure
of (31b), but instrumental zanjatijami denotes a patient rather than
an agent and genitive Ivana denotes an agent rather than a patient.
In this instance of nominalization, instrumental zanjatijami functions as an object and genitive Ivana functions as a subject.

It is not difficult to find more examples of nominalizations in
which instrumentals denote patients rather than agents and genitives
denote agents rather than patients. This type of nominalization
occurs with a large class of verbs that take an object in the instrumental, like rukovodit', 'to guide,' upravljat'to manage,' torgo-
vat' 'to sell,' etc.

All these examples show that Russian does not use any coding devices to make ergativity a distinct grammatical category in nominal
constructions. True, ergativity differs from other relations denoted
by the instrumental in Russian nominal constructions; but since
ergativity is not distinguished from other relations in the opposition instrumental:genitive by at least one coding rule, ergativity
does not constitute a distinct grammatical category and is simply a
member of the class of relations denoted by the instrumental in
Russian nominal constructions.

One may object to the above analysis of the ergative pattern and
the meaning of the instrumental in Russian nominal constructions by
pointing out that in Dyirbal and other Australian languages the instrumental is used both as an equivalent of the ergative and as the
instrumental in other ergative languages. Why, one might ask, do I
consider Dyirbal to have the grammatical category 'ergative' and
deny that Russian has this grammatical category?

My answer is that any syntactic pattern must be considered in
its relationship to the overall system of the language to which it
belongs. The syntactic patterns with the instrumental are very different in Dyirbal and in Russian. True, the instrumental merges
with the ergative in Dyirbal. But two instrumentals, one with the
meaning 'agent' and another with the meaning 'instrument,' can contrast within the same sentence in Dyirbal, which is ungrammatical in
Russian.

Consider the Dyirbal sentence:

(34)   balan ŋugumbil baŋgul yaraŋgu bangu yuguŋgu   balgan

        the woman   [by] the man   [with] the stick   is being
                                                          beaten

        Absolutive   Instrumental   Instrumental
                       (=agent)   (=instrument)

A similar sentence with agent-instrumental and instrument-instrumental is ungrammatical in Russian:

(35) *Mužčina   byl bit   palkoj   ženščinoj

Likewise, nominal phrases, such as:

(36) *Razrušenie   goroda   vragom   bombami.

The destruction(nom) of the city(gen) by the enemy(instr) with bombs (instr)

The man(nom) was beaten with the stick(instr) by the woman (instr)

Agent-instrumental and instrument-instrumental are in complementary distribution in Russian, while they contrast in Dyirbal. Besides, sentences with agent-instrumentals are basic, that is unmarked, in Dyirbal, while in Russian sentences with agent-instrumentals are passive constructions, that is non-basic, marked constructions. Actually, Russian nominal constructions with agent-instrumentals are analogues of Russian passive constructions.

The above consequence is of paramount importance for typological research: with respect to ergativity, only those syntactic processes are typologically significant which are reflected by morphological processes.

Here are some phenomena which are typologically significant for the study of ergative processes: relativization, split ergativity, extraction rules (called so because they extract a constituent from its position and move it to some other position; the term 'extraction rules' covers WH-question, relativization and focus), antipassives, possessives.

The important thing to note is that the ergative processes connected with these phenomena in ergative languages have no counterparts connected with the respective phenomena in accusative languages; they characterize only different types of ergative languages.

In treating ergativity as a grammatical category, we come across the following question: Is ergativity identical with agentivity?

Under the definition of the grammatical category proposed above, ergativity is identical with agentivity if we define the meaning 'agent' as a class of meanings characterized by the same coding as the syntactic function 'ergative.'

The above claim that agent is a grammatical category in ergative languages is opposed to the currently prevailing view that the notion

'agent' is a non-formal, purely semantic concept. Thus, Comrie
writes:

> I explicitly reject the identification of ergativity and agen-
> tivity,... despite some similarities between ergativity and
> agentivity, evidence from a wide range of ergative languages
> points against this identification.

(Comrie, 1978:356)

To support this view, Comrie quotes examples, such as the following
sentences from Basque (Comrie, 1978:357):

(37)   a.   <u>Herra</u> -<u>k</u> <u>z</u> -<u>erabiltza</u>.

           hatred-Erg you-move

           'Hatred inspires you.'

       b.   <u>Ur-handia-k</u>   <u>d</u> -<u>erabilka</u> <u>eihara</u>.

           river   -Erg it-move   mill-Abs

           'The river works the mill.'

Such examples show that agentivity is denied formal status in
ergative languages because of the confusion of the lexical and grammatical meanings of nouns in the ergative case.

Lexical meanings are meanings of morphemes which constitute word
stems, while grammatical meanings are the meanings of inflexional
morphemes, prepositions, conjunctions, and other formal devices,
such as word order. Lexical meanings are not necessarily congruous
with grammatical meanings they are combined with. There may be a
conflict between the lexical and grammatical meaning of a word. For
example, the grammatical meaning of any noun is 'thing,' but the
lexical meaning of a noun may conflict with its grammatical meaning.
Thus the lexical meanings of the word <u>table</u> or <u>dog</u> are congruous
with their grammatical meaning, but the lexical meaning of <u>rotation</u>
(process) or <u>redness</u> (property) conflict with their grammatical
meaning 'thing.' The grammatical meaning of verbs is 'process.' In
verbs like <u>to give</u> or <u>to walk</u> the lexical meanings refer to different
actions and, therefore, they are congruous with the grammatical meaning of the words. But consider the verbs <u>to father</u> or <u>to house</u>.
Here the lexical meanings conflict with the grammatical meaning of
verbs. Lexical meanings are closer to reality than grammatical
meanings. The differences between word classes are based not upon
the nature of the elements of the reality the words refer to, but
upon the way of their presentation. Thus, a noun is a name of anything presented as a thing; a verb is a name of anything presented

as a process. If we confuse lexical and grammatical meanings, we
will be unable to distinguish not only between the main classes of
words, but also between any grammatical categories. A case in point
is the grammatical category of agentivity.

From a grammatical point of view, any noun in the ergative case
means 'agent,' no matter what its lexical meaning is (that is, the
meaning of the stem of a given noun). In Comrie's examples, the
lexical meanings of herra-k in (37a) and of ur-handia-k in (37b)
conflict with the meaning of the ergative case, which is a grammatical meaning. But the crucial thing is the grammatical meaning. The
ergative case has nothing to do with the objects of reality the lexical meanings of nouns refer to. The ergative case has nothing to do
with real agents; rather, the ergative case is a formal mode of
presentation of anything as an agent, no matter whether this is
adreal agent or not. Contrary to the current view, the agent is a
formal notion in ergative languages. This claim is based on a
strict distinction between lexical and grammatical meanings.

While in ergative languages the agent is a grammatical category,
it does not have a formal status in accusative languages. In these
languages, the agent is a variant of the meaning of the nominative,
instrumental, or some other cases or prepositional phrases. For
example, we may use the term 'agent' when speaking of passive constructions, but only in the sense of a variant of some more general
grammatical category, because there are no distinct coding devices
that separate the meaning 'agent' from other related meanings, such
as the meaning 'instrument.' Thus, in English, by introduces an
agent in the passive, but can have other meanings as wel. Compare:
written by Hemingway and taken by force, earned by writing, etc.

Now, I want to propose a definition of the class of ergative
languages. In the light of the semiotic approach to ergativity, the
class of ergative languages must be defined as follows:

Ergative languages are the languages with the grammatical
categories 'agent' and 'ergative construction.'

This definition covers both pure and mixed ergative languages because it does not exclude the grammatical categories 'accusative construction' and 'patient.' And it distinguishes both pure and mixed
ergative languages as a class from accusative languages. No matter
whether an accusative languages has secondary ergative patterns or
not, it never has the grammatical category 'agent.' The essential
difference between ergative languages (whether mixed or not) and accusative languages is determined by the opposition of the grammatical
categories agent:patient.

The grammatical category 'patient' (expressed by the accusative)
is obligatory for accusative languages and optional for ergative lan-

guages, which can have this category only insofar as they have an accusative subsystem.

The grammatical category 'agent' (expressed by the ergative or its equivalents) is obligatory for ergative languages, and is incompatible with accusative languages.

Accusative languages are 'patient'-languages, and ergative languages are 'agent'-languages.

Under the proposed definition of the class of ergative languages, the central task of the typology of ergative languages is the study of coding devices denoting grammatical category 'agent.' The coding devices need not necessarily be the ergative case; they can be various oblique cases or simply word order. What is crucial, however, is the discovery of conditions under which an oblique case denotes the grammatical 'agent.' Consider, for example, the use of the instrumental in Russian and Dyirbal discussed above. I pointed out that the instrumental in the sense of 'agent' and the instrumental in the sense of 'instrument' are in complementary distribution in Russian sentences and are in syntagmatic opposition in Dyirbal sentences. The function of the instrumental and other cases must be further investigated both in Dyirbal and other languages defined as ergative. This investigation may lead to the discovery that in some languages defined as ergative, the oblique cases do not denote the grammatical 'agent' and, therefore, these languages must be redefined as non-ergative. On the other hand, we will get a deeper understanding of the conditions under which an oblique case denotes the grammatical category 'agent.'

It should be noted that the use of the ergative case in a language does not necessarily imply that this language has the grammatical category 'agent.' It is well know that some languages defined as ergative use the ergative case with intransitive verbs. This raises the question: Under which conditions is the ergative case used in a given language? If we can show that this use is determined by specific contexts, then this use must be considered as secondary to the main function of the ergative to denote the grammatical agent. If, on the other hand, the use of the ergative case with intransitive verbs is unrestricted in a given language, then the ergative case must be redefined as the nominative case and the allegedly ergative language must redefined as an accusative language.

The study of the grammatical category 'agent' involves other important problems, as, for example, the behavior of the grammatical agent under the Dominance Law or the role of the grammatical agent in various splits in case marking, but here I will not pursue the matter further.

6.2   Voices in Ergative Languages.

One important consequence of the Law Duality is that the opposition of voices in ergative languages is a mirror image of the opposition of voices in accusative languages: the <u>basic voice</u> in ergative corresponds to the <u>derived voice</u> in accusative languages, and the derived voice in ergative languages corresponds to the basic voice in accusative languages.

Since in accusative languages the basic voice is active and the derived voice is passive, this means that pure ergative languages cannot have a passive voice in the sense of accusative languages. Rather, pure ergative languages can have a voice which is converse in its effect to the passive in accusative language -the so-called <u>antipassive</u>.

The claim that the passive voice is derived from the active voice and the antipassive voice is derived from the ergative voice is supported by linguistic data across languages: passive predicates are more complex than active predicates, and antipassive predicates are more complex than ergative predicates. Thus, in many accusative languages, such as English, Russian, French, Bulgarian, Armenian, Uto-Aztec, etc., the passive predicate consists of <u>BE + past participle</u> or <u>Reflexivization affix+active predicate</u>. Various morphological processes of passive derivation are described in Keenan (1975). In very rare cases, like in Mandarin Chinese, the passive predicate does not differ morphologically from the active predicate, but in these cases passivization is characterized by syntactic coding devices. So, in Mandarin Chinese passivization is characterized by the particle bèi and a change of word order. In ergative languages, antipassive predicates are characterized by an addition of special affixes of antipassivization, for example, the affix -ɣay in Dyirbal.

It should be noted that passive constructions have a narrower range than active constructions, and antipassive constructions have a narrower range than ergative constructions, which is explained by their respective functions. Thus, the grammatical function of the passive is an omission of the agent in the sentence because it is either unknown or unimportant. Another function of the passive is stylistic; this function involves a tripartite construction in which the agent is a focus of a new information. For example, <u>This novel was written by Falkner</u> contrasts with <u>Falkner wrote this novel</u> by focusing our attention on the fact that it is Falkner who wrote the novel; in the active construction the agent is given, and in the passive construction the agent is new information. Since the two functions of the passive are rather special, active constructions have a wider range than passive ones. Similarly, the function of the antipassive is more special than the function of the ergative: if the patient is given, the sentence is ergative, and if it is new, the sentence is antipassive (Kalmar, 1979); therefore, ergative constructions have a wider range than antipassive ones.

The above linguistic data concerning the range of passive and antipassive constructions are evidence for a relation of markedness: in accordance with the Markedness Law, the passive and antipassive construction are marked members, and the active and ergative constructions are unmarked members of the markedness oppositions active: passive and ergative:antipassive.

A split ergative language can have the passive voice only as a part of its accusative subsystem.

What is called the passive voice in ergative languages by Comrie and some other linguists cannot be regarded as the true passive from a syntactic point of view. Rather, this is a construction resulting from a demotion of ergative. Thus, Comrie quotes the following sentences as an example of the passive in Basque (1978:370):

(38)   Haurra   igorria   da

         child(Abs)   send-Participle Aux-3sg

True, (38) could be translated into a passive clause in English: The child was sent. But the possibility of this translation has nothing to do with the syntactic structure of (38). Since in any transitive ergative clause absolutive means patient and ergative means agent, the demotion of ergative automatically involves the topicalization of absolutive. The crucial difference between the demotion of ergative and passivization is that passivization topicalizes the patient by means of the conversion of the predicate (John sent the child: The child was sent by John), while the demotion of ergative topicalizes the patient without any change of the predicate (see a detailed discussion of clauses with demoted ergatives in Basque in Tchekhoff (1978:88-93)). The conversion of predicates is an essential part of passivization (Desclés, Guentchéva, Shaumyan, 1985). I propose to call the constructions with demoted ergatives quasi-passive constructions.

One might argue that the word 'passive' should be used with reference to any construction involving the demotion of 'agent.' This use of the word 'passive' would, of course, cover both the constructions with converted predicates and the construction with the predicates that remain unchanged. However, the question how the word 'passive' should be used is a pure terminological issue, and involves nothing of substance. Granted that we accept the broader use of the word 'passive,' the important thing is to distinguish and not to lump together two very different types of passive constructions in ergative languages: 1) passive constructions with converted predicates involving the demotion of primary terms and the promotion of secondary terms to the position of primary terms; 2) passive constructions with only the demotion of secondary terms denoting

'agents.' The real issue is this: Can the two types of passive
constructions occur both in accusative and ergative languages? The
answer is no. The first type can occur only in accusative languages,
while the second type only in ergative languages.

Why is the first type possible only in accusative languages?
Because, in accordance with the Law of Duality, a counterpart of the
first type in ergative languages is the antipassive construction.

Why is the second type possible only in ergative languages? Because the second type involves the demotion of the secondary term
denoting 'agent.' But, in accordance with the Law of Duality, the
secondary term of a clause denotes 'agent' in ergative languages and
'patient' in accusative languages. Therefore, while the demotion of
the secondary term in an ergative construction makes it 'passive,'
the demotion of the secondary term in an accusative construction
does not make it 'passive': the accusative construction remains
active. For example, if we demote the secondary term in the active
accusative construction John is reading the book, we get John is
reading, which is again an active construction.

The above claims are deductive consequences of the Law of Duality. This law is subject to disconfirmation if counterexamples are
found which cannot be explained as deviations motivated by special
empirical conditions. The empirical study of voices in ergative
languages in order to confirm or disconfirm the Law of Duality is
one of the fascinating outcomes of the proposed theory of ergativity.

6.3   Split Ergativity.

Ergative languages tend to exhibit various splits in case markings. These splits can be explained as conditioned by the
propertiesof the ergative construction.

It is well known that in many ergative languages the ergative
construction is confined to the past tense or the perfect aspect.
How can we explain the correlation between ergative constructions
and the tense/aspect?

Since in the ergative construction the primary term denotes patient, this means that the ergative construction presents the action
from the point of view of the patient; therefore, the ergative construction focuses on the effect of the action. In the accusative
construction the primary term denotes agent, which means that the accusative construction presents the action from the point of view of
the agent; therefore, the accusative construction focuses on the
action which has not yet been accomplished. Focusing on the effect
of an action tends to correlate it with the past tense and the perfect aspect, while focusing on the action which has not yet been accomplished correlates it with the present, the future, the imperfect
and the durative aspects.

Similar explanations of the split in case marking conditioned by
tense-aspect have been already proposed by other linguists (Regamay,
1954:373; Dixon, 1979:38). What, however, has been passed unnoticed
is that accusative languages present a counterpart of this split.
Accusative languages tend to restrict the use of passive constructions to the past tense and the perfect. For example, in Old Russian
the use of passive construction was unrestricted. In Modern Russian,
however, the passive voice is confined to the past participles in
the perfective aspect. Other types of passive constructions have
been replaced by constructions with reflexive verbs which are used
as a substitute of passive constructions. The explanation of this
phenomenon suggests itself immediately if we accept the view that
the passive construction is conceptually related to the ergative
construction. Like the ergative construction, the passive construction presents the action from the point of view of the patient, and,
therefore, it tends to correlate with the past tense and the perfective aspect.

Another major split in case marking is that involving personal
pronouns. In most ergative languages nouns and personal pronouns
tend to have different patterns of case markings. For example, in
Australian languages pronouns usually have accusative case markings,
while nouns have ergative case markings (Dixon, 1976). In Caucasian
languages nouns have ergative case markings, while pronouns mostly
have an identical form for transitive agents and for patients.

The noun-pronoun split in ergative languages may be explained on
the same basis as the tense-aspect split. Since the ergative construction presents the action from the point of view of the patient,
it cannot be used in the situations where the action should be presented from the point of view of the agent, as when we use personal
pronouns (Blake, 1977).

The definition of the ergative construction proposed in this
paper provides a uniform basis for the explanation of all splits in
case markings in ergative languages.

6.4   Revision of the Class of Ergative Languages.

Languages which are represented in current linguistic literature
as ergative may be found to be non-ergative in the light of the definition of the ergative construction. Thus, Georgian is generally represented as an ergative language (or, more precisely, as a split
ergative language). For various reasons, some linguists have questioned whether Georgian is ergative at all (for example, Anderson
(1970)). In the light of the Markedness Law and the Dominance Law,
we may characterize Georgian as an ergative language which has undergone a process of the reversal of markedness. Since in Georgian the
ergative case has replaced the absolutive in intransitive clauses,

the range of the ergative case became greater than the range of absolutive, and, as a result, the two cases exchanged their places in
the markedness opposition: the ergative turned into an unmarked and
absolutive into a marked case. With the exception of some traces of
ergativity, contemporary Georgian must be considered an accusative
language.

A revision of the class of ergative languages in the light of
the proposed definition of the ergative construction may lead to an
exclusion of some other languages from this class.

7. Anticipated Results

The first benefit that may be expected from the proposed theory
of ergativity is that it will give an adequate understanding of already accumulated vast amount of facts on morphology, syntax, and
semantics of ergative languages and will set guidelines for fruitful
future field work in this domain.

The last few years have seen a significant increase in the amount
of data on ergative languages, in particular, on their syntax, but
no generally accepted solution to the problem of ergativity has yet
evolved. One might argue that the proper way to solve this problem
is an increase of field research. There is no doubt that more field
work on ergative languages is of paramount importance to deeper
understanding of ergativity. But field research cannot be fruitful
without understanding of already accumulated data and a realization
of what to look for in field research.

In testing the Correspondence Hypothesis, an empirical study of
the following syntactic properties of the ergative construction is
crucial for the corroboration of this hypothesis:

1) Omissibility of ergatives and non-omissibility of absolutives
--a cornerstone of the Correspondence Hypothesis. Although this
phenomenon is predicted by the Dominance Law, there may be deviations
due to semantic and other restrictions on the Dominance Law. Now,
we must give an exact description of these restrictions and propose
hypotheses established on independent grounds.

2) Occurrence of ergatives in intransitive clauses. Although
ergatives normally occur in transitive clauses, in some languages
they may sometimes occur in intransitive clauses. Ergatives occur
in intransitive clauses either due to special conditions (for example, when a predicate implies not only an agent denoted by the ergative but also an unspecified patient denoted by the zero sign) which
are to be investigated, or as a result of the reversal of markedness,
as in the case of Georgian.

3) The use of Equi-NP Deletion, Subject Raising and other syn-

tactic rules in ergative constructions. The current view that in
most ergative languages these rules apply only to absolutives in
intransitive clauses and to ergatives in transitive clauses is based
on inadequate empirical data. Evidence from Daghestan languages and
from other Caucasian languages suggests that these rules may be applied freely both to absolutives and ergatives, on the one hand, and
to two absolutives, on the other.

    4) The use of instrumentals and other oblique cases in ergative
constructions. A theoretical scrutiny of the data concerning the
use of oblique cases in ergative constructions may lead to a conclusion that some alleged ergative constructions are in fact some varieties of accusative constructions.

    There is strong evidence that the empirical test of the Corres-
pondence Hypothesis will be positive, confirming this hypothesis.
But I want to stress that, even if the results of the test are negative, the negative results will be no less valuable than positive
results. Since the Correspondence Hypothesis is diametrically opposed to the theories which claim that syntactic structure of ergative constructions in most ergative languages is identical with the
syntactic structure of accusative constructions, the negative results will corroborate these theories and will thus contribute to
the progress in our understanding of ergativity. [8]

    The second benefit I anticipate is a syntactic typology of erga-
tive constructions based on the notion of functional superposition.
Depending on various types of functional superposition, we will be
able to distinguish various types of ergative constructions.

    The third benefit I anticipate is that the proposed research
will call for a serious overhaul of existing theories of universal
grammar.

    In contrast to existing theories of ergativity which oppose mor-
phological, syntactic and semantic ergativity and tend to emphasize
one of these aspects at the expense of others, the Correspondence
Hypothesis underlying the proposed theory views ergativity as a unitary phenomenon which presupposes an isomorphism of morphological,
syntactic and semantic levels of ergativity.

    The opposition absolutive:ergative and subject:object are both
syntactic oppositions independent of each other. Neither ergative
constructions can be defined by accusative constructions nor accusative constructions by ergative constructions: rather, both these
types of constructions must be defined with respect to a more abstract syntactic level underlying both ergative and accusative
syntax. This abstract level is a fundamental component of AG.

    The Law of Duality reveals the interrelation of ergative and

accusative constructions in a more general semiotic framework of he
opposition of markedness, which is valid not only in syntax, but
morphology, and in general semiotics, as well.

One important consequence of the Correspondence Hypothesis is
that the morphology of an ergative language corresponds to some of
its essential syntactic properties. Only those syntactic properties can be called ergative and have typological significance which
have a counterpart in morphology.

With respect to the necessity of an overhaul of the existing theories of universal grammar, the following points are especially important:

1) Since the notions of subject and direct object are not relevant for ergative languages, they cannot be considered universal.
Therefore, they must be abandoned as primitive syntactic functions
of universal grammar and replaced by the concepts of primary and secondary terms, which are defined on the basis of the primitive concepts 'operator' and 'operand.'

2) The new theory of ergativity calls for revision of the Accessibility Hierarchy in terms of the Applicative Hierarchy.

3) Contrary to the common approach to syntax which disregards,
or at least underestimates, morphological data, the new theory of
ergativity calls for a careful study of morphological data.*

---

*I am grateful to Edith A. Moravcsik for many helpful comments on
the earlier version of the paper.

Notes

1. The Principle of Semiotic Relevance, which constitutes a basis
for proper methods of abstraction in linguistics, can be traced back
to the Principle of Abstract Relevance (<u>Prinzip der abstraktiven</u>
<u>Relevanz</u>) proposed by Karl Buhler (Bühler, 1931:22-53). Bühler's
principle was meant only as a general point of view; it did not
specify the correlation between sound differences and meaning differences as an essential condition on constraining linguistic ab-
stractions.

2. The notion of sign proposed here differs from Saussure's notion
(Saussure, 1966:65-70). Saussure considers a sign to be a bilateral
entity consisting of a signifiant and a signifié. Although Saussure
must be given credit for demonstrating that the signifiant and
signifié are intimately united, his notion of the sign lacks the
fundamental distinction between the concepts 'signifiant of' and
'signifiant,' between 'signifie of' and 'signifié' (in my termi-

nology, between 'sign of' and 'sign,' between 'meaning of' and 'meaning').

3. The notion of the grammatical meaning as obligatory in a language was proposed by Franz Boas (Boas, 1938; see also Jakobson, 1971: 488-496).

4. The notion of markedness was introduced into poetics and linguistics by Roman Jakobson under the influence of P. Verrier's work Essai sur les principes de la métrique anglaise (1909-1910). This notion was further developed by him in collaboration with Trubetzkoy. The Markedness Law had a central place in the works of the Prague School (see Jakobson, 1971a for the history on the notion of markedness). The precise statement of the Markedness Law belongs to Kuryɫowicz (see Kuryɫowicz, 1975).

5. A complete description of the formal apparatus of Applicative Grammar is given in Shaumyan, 1977; 1985.

6. The Applicative Principle was proposed by the Russian mathematician Moses Schönfinkel (see Schönfinkel, 1924).

7. If we regard terms and sentences as zero-place operators, then the rule under (20) may be restated as follows:

    1. Terms and sentences are operators.

    2. If x and y are operators, then Oxy if an operator.

8. A systematic survey of the vast theoretical literature on ergativity is outside the scope of the present paper. This is an important task in its own right which could be the subject matter of a special work.

References

Anderson, S. R. (1976). 'On the Notion of Subject in Ergative Languages'. In Li, C. N. (ed.), Subject and Topic, New York and London: Academic Press.

Aronson, Howard I. (1977). 'English as an Active Language'. Lingua 41, 206-216.

Aronson, Howard I. (1970). 'Toward a Semantic Analysis of Case and Subject in Georgian'. Lingua 25.

Blake, Berry J. (1977). Case Markings in Australian Languages Linguistic Series 23). Canberra: Australian Institute of Aboriginal Studies.

Boas, Franz (1938). 'Language'. General Anthropology. Boston.

Bühler, Karl (1931). 'Phonetik und Phonologie'. Travaux du Cercle Linguistique de Prague, 4.·

Catford, Ian C. (1975). 'Ergativity in Caucasian Languages'. Mimeograph.

Churchward, C. M. (1953). Tongan Grammar. London: Oxford University Press.
Comrie, Bernard (1979). 'Degrees of Ergativity: Some Chukchee Evidence'. In Plank, F. (ed.).
Comrie, Bernard, (1978). 'Ergativity'. In Lehman, W. P. (ed.), Syntactic Typology: Studies in the Phenomenology of Language, Austin: University Press.
Comrie, Bernard (1976). 'The Syntax and Semantics in Causative Constructions: Crosslanguage Similarities and Divergences'. In Shibatani, M. (ed.), Syntax and Semantics 6, New York and London: Academic Press. 261-312.
Curry, Haskell B. and Feys, Robert (1958). Combinatory Logic, vol. 1 Amsterdam: North Holland Publishing Company
Desclés, Jean-Pierre; Guentchéva, Zlatka; Shaumyan, Sebastian (1985). Theoretical aspects of Passivization in the Framework of Applicative Grammar. Forthcoming.
Dik, Simon C. (1978). Functional Grammar. Amsterdam: North Holland Publishing Company.
Dixon, R. M. W. (1979). 'Ergativity'. Language 55, 59-138.

Dixon, R.M. W. (1972). The Dyirbal Language of North Queensland. Cambridge: University Press.
Dixon, R. M. W. (ed.) (1976). Grammatical Categories in Australian Languages (Linguistic Series 22). Canberra: Australian Institute of Aboriginal Studies.
Einstein, Albert (1973). Ideas and Opinions. New York: Dell Publishing Co.
Harris, Alice C. (1976). Grammatical Relations in Modern Georgian. Cambridge, Massachusetts: Harvard University Press.
Jakobson, Roman. (1971). Boas' view on grammatical meaning. In Roman Jakobson, Selected Writing, Vol. 2, The Hague: Mouton, 488-496.
Jakobson, Roman (1971a). 'Krugovorot lingvističeskix terminov'. In Avanesov, P. (red.), Fonetika. Fonologija. Grammatika. Moskva: Nauka, 348-387.
Johnson, D. E. (1976). 'Ergativity in Universal Grammar'. To appear in Perlmutter, D. M. (ed.), Studies in Relational Grammar.
Kalmár, I. (1979). 'The Antipassive and Grammatical Relations in Eskimo'. In Plank, F. (ed.).
Keenan, Edward L. (1975). "Towards a Universal Definition of 'Subject'". In Li, C. N. (ed.), Subject and Topic, New York and London: Academic Press.
Keenan, Edward L. (1975). 'Some Universals of Passive in Relational Grammar'. Papers from the Eleventh Regional Meeting, Chicago Linguistic Society, University of Chicago.
Keenan, Edward L. and Comrie, Bernard (1977). 'Noun Phrase Accessibility in Universal Grammar'. Linguistic Inquiry 8, 63-69.
Kibrik, A. E. (1979). 'Canonical Ergativity and Daghestan Languages'. In Plank, F. (ed.)
Klimov, G. A. (1977). Tipologija jazykov aktivnogo stroja. Moskva: Nauka.

Klimov, G.A. (1974). 'K proisxozdeniju ergativnoj konstrukcii pred-
lozenija'. Voprosy jazykoznanija 4, 3-12.
Kuryłowicz, Jerzy (1975). "Extrapolation d'une loi linguistique".
Esquisses linguistiques II. München: Wilhelm Fink Verlag,
55-66.
Kuryłowicz, Jerzy (1960). "La construction ergative et le develop-
pement 'stadial' du language". Esquisses linguistiques I,
deuxieme edition. München: Wilhelm Fink Verlag, 95-103.
Larsen, T. W. and Norman, W. M. (1979). 'Correlates of Ergativity
in Mayan Grammar'. In Plank, F. (ed.).
Martinet, A. (1975). Studies in Functional Syntax. Munchen:
Wilhelm Fink Verlag.
Mel'čuk, Igor (1983). 'Grammatical Subject and the Problem of the
Ergative Construction in Lezgian'. Papers in Linguistics 2.
Studies in the Languages of the USSR. Edmonton: Linguistic
Research.
Moravcsik, Edith A. (1978). 'On the Distribution of Ergative and
Accusative Patterns'. Lingua 45, 233-279.
Plank, Frans (ed.) (1979). Ergativity: Towards a Theory of Gram-
matical Relations. London and New York: Academic Press.
Postal, P. M. (1977). 'Antipassive in French'. NELS 7, 273-313.
Quang, Phuc Dong (1971). 'English Sentence without Overt Grammati-
cal Subject'. In Zwicky, A. M. et al. (eds.), Studies Out in
Left Field: Defamatory Essays Presented to James McCawley,
Edmonton: Linguistic Research, 3-10.
Regamay, C. (1954). "A propos de la 'constuction ergative' en
Indo-Arien Moderne". Sprachegeschichte und Wortbedeutung.
Festschrift Albert Debrunner. Bern: Francke Verlag, 363-384.
Sapir, E. (1917). "Review of 'Het passieve Karakter van het Verbum
actionis in Talen van Noord-Amerika' by C. C. Uhlenbeck". IJAL
1, 82-86.
Saussure, Ferdinand de (1966). Course in General Linguistics.
New York: Philosophical Library.
Schmerling, Susan (1979). 'A Categorial Analysis of Dyirbal'.
Texas Linguisitic Forum, vol 13, 96-112.
Schönfinkel, M. (1924). 'Uber die Bausteine der mathematischen
Logik'. Mathematische Annalen 92, 305-316.
Schuchardt, H. (1905-1906). 'Über den aktivischen und passivischen
Charakter des Transitivs'. Indogermanische Forschungen 18, 528-
531.
Shaumyan, Sebastian (1985). Semiotic Theory of Language. Forth-
coming.
Shaumyan, Sebastian (1981). 'Constituency, Dependency, and Applica-
tive Structure'. To appear in Makkai, Adam and Melby, Alan K.
(ed.), The Rulon Wells Festschrift.
Shaumyan, Sebastian (1977). Applicative Grammar as a Semantic
Theory of Natural Language. Chicago: University of Chicago
Press.
Tchekhoff, Claude (1978). Aux Fondements de la Syntaxe: L'Ergatif.
Paris: Presses Universitaires de France.

Van Valin J., Robert D. (1977). 'Ergativity and Universality of
    Subjects'. In Beach, W. A., Fox, S. E., Philosoph, S. (eds.),
    Papers from the Thirteenth Regional Meeting, Chicago Linguistic
    Society, University of Chicago.
Vinogradov, V. (red.) (1967). Jazyki narodov SSSR, vol, 4. Iberi-
    jskokavkazskie jazyki. Moskva: Nauka.

MARKEDNESS, THE ORGANIZATION OF LINGUISTIC INFORMATION

IN SPEECH PRODUCTION, AND LANGUAGE ACQUISITION

Steven G. Lapointe

Wayne State University

In recent work (Lapointe, 1982, 1985) I have been exploring some
extensions of Garrett's (1975) model of normal adult sentence production in an attempt to explain certain aspects of agrammatic aphasia. In the present paper, I would like to consider the role of
markedness in those extensions of the normal production system and
to explore the relationship among markedness principles, the information manipulated by the speech system, and child language acquisition. The paper will proceed in the following way. First, I will
outline the relevant extensions of Garrett's model, concentrating
specifically on the operations of the Syntactic Processor, the most
important of which for the present discussion involves accessing
information from two types of stores -- one containing fragments of
syntactic structures, the other containing function words. These
stores are organized according to markedness principles which relate
structural fragments and function words to the semantic notions that
they typically express. Because it is important for the subsequent
discussion, I will describe in some detail the principles underlying
this store organization for the particular case of VP fragments, V
inflections, and auxiliary (Aux) Vs.

After summarizing these extensions of Garrett's model, I will
turn next to the issue of how children might acquire information
about V forms which is computed by the Syntactic Processor in the
production system. If the order of acquisition of linguistic elements does in fact follow from universal markedness considerations,
then we should expect the markedness principles which organize the
production stores to correctly predict the order in which children
acquire V forms. Unfortunately, the predictions made by those prin-

219

ciples do not accord with the known facts. The discrepancy seems to
stem, however, from the fact that a more general markedness principle, which is not directly relevant to the organization of V form
information in the adult production system, is the chief condition
governing the acquisition of V forms by children. Following
Williams' (1981) suggestion that markedness principles be viewed as
following from procedures of the acquisition mechanism, rather than
being theoretical primitives in themselves, I will then formulate
some preliminary procedures for the acquisition of grammatical
information concerning V forms and for the subsequent acquisition of
information in the productions stores, and I will show how the known
acquisition facts follow from these procedures. Finally, I will
comment on several of the broader issues raised in the course of
this study.

1. The Adult Syntactic Processor

1.1 Garrett's model. Garrett's original sentence production model
is summarized in (1) below.[1]

(1) Garrett's (1975) Production Model

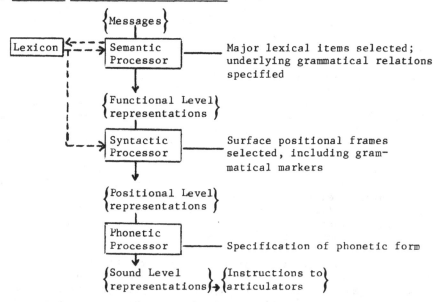

According to this model, sentences are produced in the following way.
First, the speaker begins with various Internal Messages which s/he
wants to express. On the basis of those Messages, major lexical
items are looked up in the mental lexicon, and the Semantic Processor
constructs a Functional Level representation which includes information about the underlying grammatical relations obtaining among the

major lexical items that have been selected. These representations
are then given to the Syntactic Processor which selects a positional
frame containing grammatical markers already in their surface positions on the basis of the information contained in the Functional
Level representations and inserts the phonemic representations of
the selected major lexical items into slots in these frames to yield
Positional Level representations. These are next sent to the Phonetic Processor which produces a Sound Level representation containing
all the necessary phonetic information about the utterance, and this
information is in turn converted into Articulatory Instructions.

1.2 Overview of Extensions to the Syntactic Processor

While there are several modifications which we might reasonably
consider making in this model,[2] I want to focus here on extensions
to the Syntactic Processor in (1) (henceforth, SP). The elabora-
tions that need to be considered involve a more detailed specification of the information that is computed by the SP and the operations
that actually carry out those computations. I will take up each of
these sets of issues in turn.

With respect to the information that is manipulated by the SP,
three aspects of the model in (1) must be clarified: the kinds of
information carried by the Functional Level representations that are
input to the SP, the question of what exactly constitutes a 'positional frame', and the way in which positional frames are stored in
the production system.

Regarding the information in Functional Level representations,
Garrett's discussion is not entirely clear. Although he speaks in
terms of underlying grammatical relations, it seems from his discussion that we could just as easily say that these representations
specify the linguistically relevant semantic relations obtaining
among constituents, an assumption that I will adopt below. In particular, I will assume that such relations in the domain of V forms
can be abbreviated as in (2), corresponding to the 3sing present
form in English.

(2) (indicative, active, nonspecific, present, singular-3pers)

I will refer to individual semantic notions in such representations
(e.g., 'indicative', 'present', etc.) as basic notions, and sequences
of basic notions enclosed in parentheses, as in (2), as compound
notions. The actual representations of the relations underlying
such abbreviations will be left as an open question for future
research.

Next, it is reasonable to identify Garrett's positional frames
with fragments of surface syntactic structures of the sort in (3).
Such fragments consist of maximal phrase structures of lexical cate-

gories (e.g., VP here), are expanded down to the stem-level of morphological structure for the lexical head of the phrase (here, V_s
= V stem), and indicate where other syntactic constituents are to be
attached (the circled NP in (3)) and where dependent function-words
are to be inserted (the slot under the Aux node).[3]

(3)

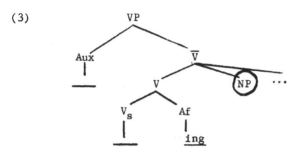

    Finally, if we take positional frames to be structures like (3),
how do they arise in the production process and what does the SP do
with them? A natural answer to these questions is that there are
two stores that interact specifically with the SP in production, one
containing fragments like (3) and the other containing dependent
function words like the auxiliaries, and that part of the process of
producing a Positional Level representation involves the SP's accessing fragments and function words from these stores and combining
them in the grammatically appropriate ways. If this view is correct,
then an important question to ask is, How are these stores organized?
From work on agrammatism (Lapointe, 1982, 1985) it appears that the
stores are organized basically in terms of markedness principles
which govern the compound semantic notions that the various
fragments and function words express. In attempting to state these
principles explicitly, however, one must worry about a lot of details
since this is a particularly complex and poorly understood area of
linguistic research. Recognizing these problems, I will leave these
issues for the moment and will return to them shortly in a separate
subsection below.

    Returning now to the operations carried out by the extended SP,
it is clear that that component must perform the following operations: (a) it must accept information about Functional Level representations, including information about semantic notions like
those abbreviated in (2), (b) it must decide on the basis of that
information which fragments and function words are required, (c) it
must actually locate and copy out fragments and function words from
their stores, (d) it must combine fragments together and add the
function words where they are required, and (e) it must insert
phonemic information about the major lexical items that have already
been selected. In work reported elsewhere (Lapointe, forthcoming),
I have argued that the grammatical marker deficits found in agrammatism result from damage to the mechanism specifically responsible

for accessing information from the fragment and function word stores.
The particular claims made in that work allow one to account for a
wide range of facts regarding V marker agrammatism in English and
Italian, and to connect this deficit with the syntactic structure
disturbances that usually accompany grammatical marker impairments.
However, since the specific manner in which these operations are
performed by the SP, in particular, the way in which the fragment
and function word stores are accessed, does not directly bear on the
discussion below, I will say nothing further here about these
operations, inviting the interested reader instead to consult the
work just cited for more details.

1.3 Organization of the production stores

Having outlined the above extensions to the SP in (1), I would
now like to return to a more detailed examination of the organization of the fragment and function word stores. Before formulating
the required organization principles themselves, I will first state
a few definitions and consider some specific examples.

To begin, let us adopt the extended definition of <u>derivation</u> for
fragments given in (4).

(4) <u>Def.</u> A V fragment G is said to be derived from <u>base</u> fragment F
if one of the following holds.

a. G contains a single V form and F contains the single V form
whose stem is used to derive the form in G via the standard
inflectional rules of the language

b. G contains an Aux+V sequence, F contains either a single V
form or an Aux+V sequence, and G is formed by adding an
Aux to F under the syntactic constraints of the language.

In the specific case of V fragment stores, three types of fragments
occur: derived and base fragments containing single V forms of the
sorts defined in (4a), examples of which are given in (5) below, and
base fragments of the sort defined in (4b), an example of which is
given in (3) above.

(5) a. A derived fragment   b. A base fragment
defined by (4a)   defined by (4a)

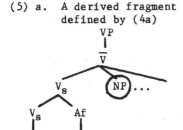
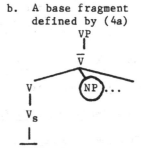

To conserve space, I will henceforth abbreviate fragments like
(5a,b) and (3) as in (6).

(6) a. V+$\underline{s}$ (=5a))
    b. V   (=5b))
    c. Aux V+$\underline{ing}$ (=(3)

Turning next to the compound semantic notions expressed by such
fragments, we can identify two kinds as in (7) depending on the hierarchies of basic notions in (8).[4]

(7) Def. A compound notions is said to be complete if it contains at
least one notion from each of the hierarchies in (8) that is expressed in the particular language; otherwise, it is said to be
incomplete.

(8) Markedness orderings for basic V notions

     a. Attitude: Indicative < nonindicative (modal, subjunctive,
                                          imperative, etc.)

     b. Voice: Active < passive < causative
     c. Aspect: Nonspecific < durative < punctual, completive

     d. Tense: Present < past, future < complex < nonfinite
     e. Agreement: Subject < direct object < ...; singular < plural <
                        dual < ...; 3pers < 2 pers < 1 pers

Various conditions govern the legal combinations of the basic notions
in (8) into compound notions like the one given in (2). Investigation
of such conditions has formed the traditional domain of research for
comparative and historical morphology, but I will not undertake a
detailed study of these conditions here.

In general, the following relations seem to hold between the
types of fragments defined in (4) and the types of notions defined
in (7). Derived fragments generally express complete compound notions, base fragments generally express incomplete notions, and the
notions expressed by base fragments are generally a subset of those
expressed by derived fragments, where in the process of derivation
defined in (4) the missing basic notions in the compound notion expressed by a base fragment are added to form the complete compound
notion expressed by the derived fragment.

Now, it is possible to rank complete compound notions according
to the principle in (9) if we assume that the hierarchies in (8) are
themselves ranked from top to bottom in the order given there.

(9) Markedness rankings for complete compound notions

     Two complete compound notions are ordered according to the

rankings given by the highest scale in (8) on which they differ
in basic notions.

To take an example, consider the complete compound notions P and Q
in (10).

(10)   P = (indicative, active, durative, past, subj-pl)
        Q = (indicative, active, punctual, present, subj-sing)

Here P < Q because the highest scale on which they differ is the
Aspect hierarchy (8c), and durative < punctual there. The ranking of
incomplete compound notions is a far more complex matter, but as we
are about to see, the details of that ordering are not crucial for
the discussion.

   We are nearly ready to organize V fragments in their production
stores; however, we must first take care of three further issues.
First, a given fragment sometimes expresses several distinct compound
notions. When this occurs and the fragment is derived, it seems
reasonable to assume that the least marked complete compound notion
expressed by the fragment determines the ranking of that fragment.
Second, the ranking of base fragments like (6b,c) is reasonably determined by the least marked fragments derived from them. Finally,
where the ranking of a base fragment for a single V form has not
already been uniquely determined, it is placed immediately before
the fragments which it is used to derive. These issues can be handled by first associating each fragment with a primary compound
notion as in the definition in (11), and then by defining the organization of V fragment stores in terms of primary notions as in (12).

(11)   Def. The primary notion of a derived fragment F is the
        least complex complete compound notion expressed by F. The
        primary notion of a base fragment F is the least complex
        complete compound notion expressed by a fragment G derived
        from F. In the latter case, G is referred to as the primary
        fragment of F.

(12)   Organization of V fragment store

        a. Group together in columns all fragments whose primary
            notions express the same major verbal notions (Atti-
            tude, Voice, Aspect), with columns ordered according
            to (8) and (9).

        b. Within columns arrange fragments by the Tense and
            Agreement notions expressed by the primary notions of
            the fragments according to (8) and (9).

        c. Order base fragments for single V forms immediately
            before their primary fragment.

Some example base and derived fragments in English and some of the
notions they express are given in (13) and (14).[5]

(13) Some English base V fragments

    a. V -- (indic., act., nonspec.)
        (indic., act., nonspec., pres., sing-2)

           .
           .
           .

    b. Aux V+ed' -- (indic., act., nonspec., anter.-X)
    c. Aux V+ing -- (indic., act., durat.)
    d. Aux V+ed" -- (indic., pass., nonspec.)
    e. Aux V   -- (nonindic-X, act., nonspec.)

    Some English derived V fragments

(14) a. V+s -- (see (2) above)
    b. V+ed -- (indic., act., nonspec., past)
    c. has V+ed' -- (indic., act., nonspec., anter-pres, sing-3)
    d. is V+ing -- (indic., act., durat., pres., sing-3)
    e. is V+ed" -- (indic., pass., nonspec., pres., sing-3)
    f. will V -- (nonindic.-future, act., nonspec., pres.)

The V fragment store for English based on these forms and notions is
given in partial form in (15), where the first column includes fragments expressing (indicative, active, nonspecific), the second column
includes fragments expressing (indicative, active, durative), and so
on, in accordance with (12a).

(15) V fragment store for English (partial)

V	Aux V+ing	Aux V+ed"	Aux being V+ed"	Aux V ...
V+s	Aux been V+ing	Aux been V+ed"	Aux been being V+ed"	Aux have... V+ed'
V+ed				
Aux V+ed'				

For the past several pages, we have been concerned exclusively
with the organization of V fragment stores. Now we must ask how the
Aux stores are organized. Unfortunately, the answer to this question
is a good deal less clear at this point. Here, I will simply assume
that Aux stores have columns consisting of particular forms of separate Aux's, with the forms within the columns determined by the Tense
and Agreement notions that the forms express. In English, it seems
that the columns are ordered with forms of be first, then forms of
have, then forms of do, and finally forms of the modals; although

this seems to be the correct order, it is not at all obvious which of
several plausible alternative principles force this ordering of the
Aux's.

2. Acquisition -- A First Pass

To summarize briefly, we have accomplished the following so far.
First, we have considered Garrett's production model in (1), focusing specifically on the SP. Secondly, we have identified positional
frames in that original model with structural fragments like (3),
(5a), and (5b). Third, on the basis of earlier work on agrammatism,
we have found it useful to assume that these fragments are organized
in stores that are consulted by the SP during processing according
to the markedness of the semantic notions that the fragments express.
Fourth, we have seen that the relevant principles for V fragment
stores are the ones given in (12), yielding a store like (15) for
English.

It is now reasonable to ask how the child might acquire the information found in the fragment and function word stores. After
all, the idea that markedness principles are related to acquisition
has been around at least since the appearance of Jakobson's (1968)
work, and the claim has been made that the markedness principles in
(12) are used to organize these production stores. So, perhaps
these principles govern the acquisition of V forms in child language.
If this is correct, given what has been said so far, we would expect
the child first to acquire the form corresponding to the fragment in
the first cell of the first column, then to acquire the forms corresponding to each of the remaining fragments in the first column in
turn, one at time, along with any accompanying Aux's where applicable, then, when the child is finished with the forms from the first
column, to go to the top of the next column to the right and to continue in the same manner until all the forms have been learned.

Such a procedure would lead to the expected acquisition order in
(16). Unfortunately, there are several discrepancies between this
expected order and what seems to be the actual order in (17), based
on work by Brown (1973) and Bellugi (1967).

(16) <u>Acquisition order expected under (12), (8), (9)</u>

   V<V+<u>s</u> < V+<u>ed</u> < <u>have</u> V+<u>ed</u>'< <u>be</u> V+<u>ing</u><<u>have been</u> V+ing<...<modal V<...

(17) <u>Actual acquisition order</u> (based on Brown, 1973; Bellugi, 1967)

                  V+<u>s</u>
   V <V+<u>ing</u>   <   V+<u>ed</u>   < <u>be</u> V+<u>ing</u> << <u>have</u> V+<u>ed</u>', <u>be</u> V+<u>ed</u>"< ...
                  modal V

V+<u>ing</u> is present in the actual order of acquisition and is in fact
learned before the simple present and past tense forms, although the

form is completely absent in the expected order in (16). Modals
appear to be learned at about the same time that the simple present
and past tense forms are acquired, even though this is anomalously
early according to (16). Progressive forms (be V+ing) are learned
before perfect forms (have V+ed'), contrary to the expectations
summarized in (16), and perfect and passive forms are learned much
later than is to be expected if (16) were correct.

The problem now is to try to explain the differences between (16)
and (17). To solve this problem, I do not want to abandon the assumption that there is a relation between markedness and acquisition;
rather, I would like to argue that there is more to the markedness
story than we have considered up to this point, and that by looking
into the matter a little more closely, we will discover the clues
needed for solving the problem.

What we have been overlooking so far is the basic and very general principle often assumed to govern the relation between forms of
major lexical items and the semantic notions those forms expresss.
This principle is given in (18).

(18)   Fundamental Markedness Principle Relating Forms and Notions

        A grammar is less marked to the extent that less complex mor-
        phosyntactic forms express less complex compound semantic
        notions.

The reason why we have overlooked this principle is that it is irrelevant as far as the adult production system is concerned. That system only cares about whatever actual form/notion pairings exist in
the adult language, and, as we have seen, fragment and function word
stores can be organized on that basis alone. On the other hand, the
principle in (18) makes claims about which form/notion pairings are
to be expected. Hence, from the perspective of the child language
learner, (18) would be a handy tool to employ in trying to learn
grammatical markers in the first place. If this view is correct,
and the child is using something like (18), then s/he will have to
be concerned simultaneously with both the complexity of the semantic
notions involved and the complexity of the forms expressing those
notions.

In the remainder of this paper, I want to explore in more detail
how the child juggles these two types of information in acquiring V
forms. In doing so, I will adopt Williams' (1981) approach to explaining markedness phenomena in acquisition, which is somewhat
different from the more usual Jakobsonian notion. In Williams'
approach, markedness principles like (18) are assumed to play no
theoretical role per se; instead, theoretical status is accorded to
procedures of the acquisition device, from which the markedness principles follow. Thus, markedness principles are viewed as analogous

to physical laws, like the ideal gas laws of nineteeth century
physics -- they themselves are not assumed to be fundamental statements of the theory but are seen instead as signposts along the road
in search of the more basic theoretical principles which they follow
from.

3.   Acquisition Procedures for V Forms in the Grammar

In formulating these procedures, I will first investigate the
grammatical considerations involved in the acquisition of V forms
and Aux V sequences, and then I will return to consider the way in
which the acquisition of these forms leads to the acquisition of
fragments in the production stores. I will be claiming that the
basic determiner of acquisition here involves the complexity of V
forms, with the acquisition of semantic notions dependent on the
acquisition of progressively more complex forms.

Taking up the issue of V form complexity first, two factors seem
to be involved in determining the complexity of a fragment in terms
of formal properties: (a) the number of Aux's present, and (b) the
number of affixes (Af's) attached to the V, if the V is inflected.
We can ignore the number of Af's attached to the Aux's, since Aux's
are generally irregular in form, and so we can assume that the child
simply has to learn each of the particular Aux forms separately.
These two factors can be combined to give the complexity principle
for forms in (19), which can be represented by the matrix in (20).

(19)   Complexity rankings for V forms

   a.   Rank together V groups on the basis of the number of Aux's
        they contain.

   b.   Within sets of V groups containing the same number of Aux's,
        rank V groups on the basis of the number of affixes (Af's)
        attached to the Vs in the groups.

(20) Matrix for V form complexity

V	Aux V	Aux^2 V	Aux^3 V	Aux^4 V	...
V+Af	Aux V+Af	Aux^2 V+Af	Aux^3 V+Af	Aux^4 V+Af	...
V+Af^2	Aux V+Af^2	Aux^2 V+Af^2	Aux^3 V+Af^2	Aux^4 V+Af^2	...
V+Af^3	Aux V+Af^3	ux^2 V+Af^3	Aux^3 V+Af^3	Aux^4 V+Af^3	...
.	.	.	.	.	
.	.	.	.	.	
.	.	.	.	.	

In European languages, it is generally not necessary to go below the
second row in (20) which has been marked with the dotted line. This

constraint will result in an effective ordering of fragments by their
forms as in (21). As can be seen, this order looks a lot more like
the actual acquisition order in (17) than the earlier predicted order
(16) did, suggesting that this approach is indeed on the right
track.

(21) $V < V+Af < Aux\ V < Aux\ V+Af < Aux^2\ V < \ldots$

The procedures governing the acquisition of V forms can now be formulated to first approximation as in (22).[6]

(22)   <u>Procedures for acquiring the grammatical aspects of V fragments</u>

    a. Begin with the least complex V form defined by (19).

    b. Seek evidence for the existence of specific V forms within
        the type currently being considered. Posit those forms
        for which evidence is found.

    c. For a V form involving the addition of an Aux, check back
        over the already posited V forms to determine whether there
        are any forms X which were misanalyzed and should in fact
        be included in the Aux X type now being considered. For
        such V forms, eliminate the old groups and go back to
        step (b).

    d. Go to the next most complex V form defined by (19) and return
        to step (b).

Now we must consider how the complexity of V semantic notions is
to be incorporated into the procedures in (22). The problem that
arises in learning semantic notions is that the child has to learn
which of the universally possible notions that <u>can</u> be expressed by a
V form are in fact expressed by the V forms in the particular language being learned. This problem is therefore parallel to that of
having to learn the phonetic contrasts actually exhibited in the
sound inventory of the particular language at hand.

Faced with this problem, the child apparently proceeds in the
following way. When the child finds a new candidate V form under
step (22b), he tries to construct a compound semantic notion for
that form on the basis of the contrasts among the basic notions in
(8) which s/he has already acquired. The child then searches for
semantic evidence that this V form expresses a notion from a <u>new</u>
basic notion contrast; in doing this, the child looks for evidence
of the different types of basic notions in the following order:
Aspect notions first, then Tense and Agreement notions, and finally
the other major verbal notions (Attitude and Voice). That is, in
searching for new contrasts, it appears that the child begins with a
<u>middle-level</u> type of basic notion and then works his/her way outward,
in much the same way that the child appears to start with <u>middle-</u>

level categories and works outward in the acquisition of nouns
(Anglin, 1977). After successfully constructing a compound notion
to associate with the new V form, the child posits the new form, the
compound notion it expresses, and any new basic notion contrasts
uncovered in the process. These procedures can be included in (22)
by adopting the modifications in (23). The only portion of these
procedures that has not already been discussed is step (b.4) which
provides for the possibility that already posited compound notions
for V forms can be changed whenever a new notion contrast is posited.

(23) Modification to (22)

      Change (b) to b')
      (b') Seek formal evidence for the existence of specific V forms
           within the type currently being considered.

      Add the following under (b')
      (b.1) For each such V form G, seek semantic evidence that G
            expresses compound notions made up solely of basic
            notions from the notion contrasts that have already been
            posited.

      (b.2) Seek evidence that G expresses basic notions from (8)
            not already been posited. Consider notions from the
            following types in the order given:

               (i) Aspect notions
              (ii) Tense and Agreement notions
             (iii) remaining major verbal notions (Attitude and Voice)

      (b.3) Construct a compound notion for the V form from the basic
            notions identified in (b.1) and (b.2). Posit the new V
            form, the compound notion it expresses, and any new
            semantic contrasts discovered in (b.2).

      (b.4) Go back to already posited V forms and seek semantic
            evidence concerning whether any of them express seman-
            tic notions from the newly posited contrasts. If so,
            add the new notion(s) to all appropriate old compound
            notions.

   We are now in a position to explain the acquisition order in
(17), the various stages of which are outlined in Table 1.[7] Here
is how the child proceeds through these stages.

(A) By step (22), the child begins with single uninflected Vs, the
first type of V form in (19). There are plenty of these forms in
English, although it may not be obvious to the child exactly what
semantic notion(s) to associate with these forms. Assuming that the
child does assign some notion to these forms, it will be the least

Table 1.   Summary of the stages in the acquisition of grammatical
           information about V forms in English.

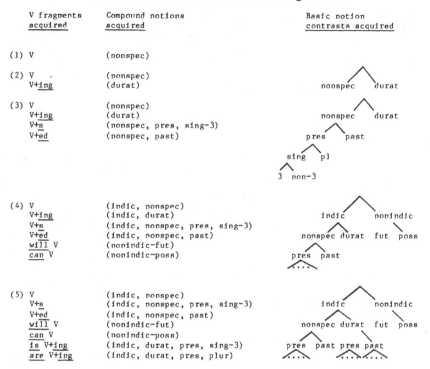

	V fragments acquired	Compound notions acquired	Basic notion contrasts acquired

marked Aspect notion in (8c), namely, nonspecific, as in stage 1 of
Table 1.

(B)   The next type of V form the child considers is the V+Af type.
Here, there are several distinct V forms that the child will encounter, none of which express only nonspecific Aspect.   Moving down
to (23b.2), the child tries to find a form which contrasts in Aspect.
Such a form is V+ing which expresses durative Aspect, in contrast to
the original V type which expresses nonspecific Aspect.   Since these
forms do not express any other basic notions, the child posits the V
forms and associated compound notions, along with the nonspecific/-
durative contrast, in his/her grammar, as shown in stage 2 of Table
1.

(C)   There are, however other V+Af forms, namely, V+s and V+ed which
the child will encounter.   These do not express just nonspecific or
durative Aspect as the other forms do, and so the child moves on to
step (23b.2) again.   No new Aspect notions are expressed by these
forms, so the child checks to see if they express any new Tense and

Agreement notions, and s/he quickly discovers that they do. As a
result, the child adds the two new forms, their compound notions,
and the new contrasts under <u>nonspecific</u> Aspect as shown in stage 3
of the table. Notice that new basic notions do not automatically
generalize to old superordinate contrasts: i.e., the newly added
<u>present</u>, <u>past</u>, etc. contrasts are not automatically added under
<u>durative</u> in the contrast tree. The reason for this seems to be that
the child needs specific evidence in order to posit a new contrast
under an already existing one.

(D) The next type of V form the child tries to find is the Aux V
type. The child does indeed find examples of such forms involving
modals plus bare V stems. Turning to the semantic notions these
forms express, the child discovers that
they do not obviously express any of the already posited notions, but
they <u>do</u> express <u>nonindicative</u> Attitudes. This leads to the additions
in the grammar shown in stage 4 of the table. Notice once again that
all of the contrasts under the <u>indicative</u> branch of the contrast tree
are not automatically generalized under the <u>nonindicative</u> branch,
again apparently for the reason mentioned in the preceding paragraph.

(E) The next form that the child must try is the Aux V+Af form.
Here the child will find frequent cases like <u>is</u>/<u>are</u>/<u>was</u> V+<u>ing</u> and
will quickly discover that s/he is in the situation described in
(22c); these V forms express <u>durative</u>, just as the V+<u>ing</u> form does,
in addition to various Tense and Agreement notions, and there is
little evidence that V+<u>ing</u> appears by itself. So, V+<u>ing</u> is eliminated in favor of the <u>be</u> V+<u>ing</u> forms, new Tense/Agreement contrasts
are added under <u>durative</u> Aspect, and the new forms and compound
notions associated with them are added to the grammar as in stage 5
of the table.

(F) The child should now also acquire the simple perfect and pas-
sive forms (<u>have</u> V+<u>ed</u>' and <u>be</u> V+<u>ed</u>") since these also involve Aux
V+Af structure, and indeed according to (17) these forms are learned
next. The only problem is that there seems to be a considerable
time lag before these forms are acquired. Several factors beyond
the control of the procedures in (22, 23) appear to contribute to
this lag. The perfect forms express a fairly complex tense notion
(<u>anterior</u>), the passive forms require manipulations of the argument
structure of the Vs in question, and both sorts of forms involve
special morphological and semantic government relations between the
Aux's and the Af's appearing in them. Because the child must master
all of this information before s/he can successfully apply the procedures (22, 23), it is reasonable to expect some time delay before
the simple perfect and passive forms are acquired. Once these forms
have been appropriately learned and the various contrasts have been
added to the child's grammar, s/he can go on to consider Aux^2 V,
Aux^2 V+Af, and the remaining V forms and notions of the language.

4.   Procedures for Acquiring V Fragments and Aux's in Production

Having sketched how the complex interactions involved in acquiring V forms and notions in the grammar yield the observed order of
acquisition in (17), we are now left with the problem of explaining
how this information finds its way into the fragment and function
word stores in the production system in such a way that we are left
with the organization outlined in section 1 above. As suggested
before, while the acquisition of V forms may be determined by (19)
and (22,23) as far as the grammar is concerned, the production
system does not care a great deal about any of this. All it wants
to see is V fragments arranged in stores according to (12) and Aux's
arranged in their stores according to whatever organization principles turn out to be the correct ones. Given the system that we have
already established, however, we can accomplish this goal fairly
easily simply by having the processing system refer to the changes
produced in the grammar after each successful application of the
procedures (22, 23). This can be acccomplished by adopting the procedure in (24).

(24)   Procedures for acquiring V fragments and Aux's in production
        stores

        a. Whenever a new major verbal notion contrast is added to
            the grammar, add new columns to the fragment store for
            the new major verbal notions. Order the new columns
            according to (8) and (9). Add the V fragments corres-
            ponding to the newly posited V forms at the head of
            the appropriate new columns.

        b. Whenever a new Tense or Agreement notion contrast is added
            to the grammar, the V fragment corresponding to the newly
            posited V form is added to the appropriate column in the
            store in the position determined by (8) and (9).

        c. Whenever a V form is eliminated and a new one added as a
            result of (22c), or changes are made to the compound
            notions of old V forms under (23b.4), the corresponding
            changes are made in the V fragment and Aux stores.

        d. Whenever a new Aux is added to the grammar as part of a
            new V form, that Aux is added in the appropriate place
            in the Aux store.

Just as we assumed that (22, 23) constitute the real theoretical
entities from which (18) follows, so we may also view (24) as the
basic theoretical object which (12) is a consequence of.[8]

Under (24) then, a child learning English V forms will behave in
the following way (cf. (24), Table 1, and (15)). At stage 1 in the

table, there is only one column in the V fragment store, and that
column contains just the fragment corresponding to the single uninflected V form. At stage 2, a second column is added to the right
of the first column corresponding to the contrast now posited between
nonspecific and durative Aspect (cf. the contrast tree at stage 2 in
the table). This column also contains only one fragment, the one
corresponding to the V+ing form. At stage 3, fragments for V+s and
V+ed are added under the fragment for V in column 1, since these do
not differ from V in the major verbal notions they express. At
stage 4, a new column is added to the right of the two existing
columns, containing fragments for Aux V forms; modal Aux's are added
as the first column in the Aux store. At stage 5, the V+ing fragment
is replaced by the Aux V+ing fragment; forms of be are added in a
new column to the left of the existing modal column in the Aux store.
Finally, when the perfect forms are learned, an Aux V+ed' fragment
is added in the first column under the existing fragments, and forms
of have are added in a new column between the columns for be and the
modals in the Aux store; when passive forms are learned. a new
column is added between the ones containing Aux V+ing and Aux V to
accomodate the new passive fragment Aux V+ed". The resulting V
fragment and Aux stores are quickly converging on the adult fragment
store in (15) and the Aux store described in section 1.1.

5.   Conclusion

It is now time to take stock of exactly where we have gotten ourselves. In an attempt to formulate an acquisition procedure which
the markedness principles in (18) will follow from, we have been led
to assume that the child has available to it a number of tools.
First and foremost, the child makes use of procedures (22, 23). In
addition, however, the child must rely on a considerable amount of
further information in order for those procedures to operate properly. Among this additional information are the following: (a)
the complexity ranking in (19) for V forms, (b) routines for the
formal segmentation of Aux's, Af's, and Vs as distinct linguistic
elements, (c) specification of what counts as "evidence" in (23b.1),
(d) some indication of the range of possible V basic semantic notions, and (e) the complexity rankings (8) and (9) for compound and
basic notions. Items (b) and (c) are likely to be specified independently of the special considerations involved in the learning of
V forms as general criteria used throughout the acquisition mechanism. Items (a), (d), and (e) on the other hand seem to be specific
to the problem of acquiring just those forms.

Furthermore, it is clear that (22, 23) are by no means in their
final forms. These procedures are deficient in at least two ways.
First, no indications are given concerning the conditions under
which the procedures stop; yet it is clear that some provisions for
this must he made in the final version of the procedures. In addition, the possibility of going back to an earlier V form in (19)

that has already been passed by (22, 23) is not allowed, but this restriction is likely to turn out to be be incorrect. Consider a language which has many single tensed V+Af forms. As stated above, the procedures force the child to learn <u>all</u> of the particular forms of this type before continuing on to consider any of the more complex types involving Aux's. It is, however, much more probable that children only learn <u>some</u> of the V+Af forms in their language at first, and then after considering more complex forms containing Aux's for a while, they return to learn some of the V+Af forms which express more complex notions. Provisions allowing for this possibility will also have to be incorporated into the final version of these procedures.

Nonetheless, despite these problems we have seen that (22, 23) can in fact account for the observed order of acquisition in (17) in a fairly natural way. Having done that, we have also seen that all of the information referred to by the organization principle in (12) for V fragment stores in production is manipulated by procedures (22, 23), and as a result, it is possible to eliminate (12) in favor of the acquisition procedures in (24).

Finally, I would like to end by drawing several broad conclusions from the work just discussed. First, it seems possible to extend Williams' initial approach to the explanation of markedness phenomena in terms of acquisition procedures in an interesting way. Williams was concerned exclusively with the problem of markedness and acquisition procedures in the case of a particular grammatical domain, that of typological correlations in the phrase structure properties among different phrase categories across grammars. I have presented a further preliminary grammatical application of Williams' approach in the treatment of the acquisition of V forms by procedures (22, 23) above. However, it now appears possible to say that markedness principles involved in organizing information <u>in the speech processing system</u> as well as in the grammar follow from acquisition procedures. Second, it is clear, however, that <u>not all</u> markedness principles can be eliminated in favor of acquisition procedures in this way. Apparently, it is just the <u>general</u> type of markedness principles, like (12) and (18), which <u>relate classes</u> of linguistic elements, that can be eliminated in this way. Principles like (8), (9), and (19) which define relative complexity among the specific members of large sets of linguistic objects need to be retained in the acquisition system in some form, since the acquisition procedures need to be able to refer to the complexities which the rankings define.

Lastly, even though they manipulate basically the same information, the procedures used to acquire information in the grammar and in the processor are slightly different, manipulating and storing the information in slightly different ways. I do not believe that this difference arises from the fact that in the above analysis, the processsing procedure (24) operates off of the information that is output from the grammatical procedures in (22, 23); it is entirely possible to construct a single set of procedures which will simultaneously posit information in the grammar and in the production stores. Even in such a system, however, the changes being affected in the grammar and in the production system being constructed by the single set of procedures will be slightly different. Instead, the reason for the close but not exact fit between the grammar and the processor in this regard seems to me to involve the distinctive burden placed on the processor. That burden, in the case at hand, is to produce grammatically well-formed speech on-line, given the limited information and resources made available to it by other parts of the cognitive system. These requirements lead to the need, among others, of having separate fragment and function word stores organized according to the semantic principles already mentioned, and these in turn seem to lead to the difference between (22, 23) and (24).

Footnotes

*The present paper, which grew out of research conducted as part of the Johns Hopkins University Aphasia Research Project, is a preliminary report on a continuing study into the properties and interactions of markedness, speech production, aphasia, and language acquisition. A longer article that incorporates the insights of this paper, along with a more comprehensive treatment of the work of Brown (1973), deVilliers and deVilliers (1973) and deVilliers (1974) on the acquisition and impairment of grammatical markers, is in preparation.

[1]As the model is usually presented, explicit mention of the lexicon and the actual processes that carry out the production computations, represented by the boxes in (1), are generally ignored, and instead the discussion focuses on the representation levels output by those procedures, represented by the braces in (1). I have included the processes here for explicitness and to avoid the confusion that often arises regarding where in this model computations are actually carried out.

[2]One aspect of this model which we might consider changing, for instance, is the seral organization of the processing components. It is likely that an organization in which information is computed in a more nearly parallel fashion will prove to be more efficient for human production. For some discussion, see Lapointe (in preparation).

[3]Note that I am only concerned with the treatment of regularly inflected V forms here. Irregular forms appear to behave rather differently. For some discussion of the general issues involved in the grammatical generation of such forms, see Lapointe (1981).

[4]The scales in (8) are intended to be representative, not exhaustive. For further discussion, see Lapointe (forthcoming).

[5]As stated, (12) says nothing about cases in which a fragment behaves like a base fragment when it expresses incomplete notions and like a derived fragment when it expresses complete notions. To handle such cases, it seems necessary to add a condition which insures that a fragment is placed in the first position it can appear in a fragment store.

In the case of certain complex basic notions, like anterior-X, a completely specified notion is substituted for the variable notion when an Aux is added to produce a derived fragment. So, anterior-X in (13b) is replaced by anterior-pres when has is added to the fragment in (14c).

In (13), (14), and (15), V+ed = past tense, V+ed' = past participle, and V+ed" = passive participle. These three forms need to be kept distinct because separate base fragments for the perfect tense and passive voice forms are required in English to allow for the derivation of distinct derived Aux V fragments.

[6]The procedure in (22) is different from the one presented in the conference in that the following step was originally included between steps (c) and (d):

(i)   if it appears that special morphological constraints, either formal or semantic, govern the appearance of Aux's and Af's in a V group, delay positing that V group until all of the constraints have been sorted out.

In a private discussion with Larry Solan, I became convinced that (i) was incorrectly stated as part of the procedures for learning V forms and that the facts that I was attempting to handle using (i), namely, the lateness in the appearance of perfect and passive forms, are to be treated more appropriately in the ways discussed below.

[7]While it is convenient to represent the semantic contrasts that the child has learned at a given stage in terms of branching tree diagrams of the sort on the right in Table 1, I do not wish to be taken as claiming that such contrast trees are actually part of the grammar. I have included these diagrams mainly as a conceptual aid to help the reader to see which of the basic notions in (8) have been acquired at a given point, but also the changes that occur from stage to stage in these contrast trees parallel in a visually clear way the changes in the V fragment and Aux stores to be discussed in the next section.

[8]However, (12) and (24) are not strictly equivalent. Consider

for example a language with a simple tensed V+Af form, distinct from the infinitive, which is used as a stem for deriving a set of regular, tensed V+Af forms. Under (12), this form would appear immediately before the forms it is used to derive; under (24), it would not. It is an open question whether this difference really matters.

References

Anglin, J. (1977) _Word, Object, and Conceptual Development_, Norton, New York.

Bellugi, U. (1967) _The Acquisition of Negation_, unpublished doctoral dissertation, Harvard University, Cambridge, MA.

Brown, R. (1973) _A First Language_, Harvard University Press, Cambridge, MA.

deVilliers, J. and P. deVilliers (1973) "A Cross-sectional Study of the Acquisition of Grammatical Morphemes," _Journal of Psycholinguistic Research_ 2, 267-278.

deVilliers, J. (1974) "Quantitative Aspects of Agrammatism in Aphasia," _Cortex_ 10, 36-54.

Garrett, M. (1975) "The Analysis of Sentence Production," In G. Bower, ed., _The Psychology of Learning and Motivation_, Vol. 9, Academic Press, New York.

Jakobson, R. (1968) _Child Language, Aphasia, and Phonological Universals_, Mouton and Co., The Hague.

Lapointe, S. (1981) "The Representation of Inflectional Morphology within the Lexicon," in V. Burke and J. Pustejosky, eds., _Proceedings of the XIth NELS Meeting_, GLSA, University of Massachusetts, Amherst, MA.

LaPointe, S. 1982. Restrictions on the use of V forms by agrammatic aphasics. Paper presented at the Annual Meeting of the Academy of Aphasia, New Paltz, New York.

LaPointe, S. 1985. A theory of verb form use in the speech of agrammatic aphasics. _Brain and Language_. 24.100-155.

Lapointe, S. (in preparation) "Towards a Theory of Speech Production Mechanisms."

Williams, E. (1981) "$\bar{X}$ Features," in S. Tavakolian, ed., _Language Acquisition and Linguistic Theory_, MIT Press, Cambridge, MA.

LANGUAGE ACQUISITION, APHASIA, AND PHONOTACTIC UNIVERSALS

Lise Menn

Aphasia Research Center
Boston University School of Medicine

In this paper I will consider the nature of phonotactic markedness in the light of work in early first language phonotaxis. Then
we will turn to work in progress on a particular family of aphasic
word production errors, to recent studies of second language acquisition, and to instrumental phonetic work on perceptual aspects of
phonotaxis. I think that all four of these areas of study, as well
as slip of the tongue data, are fitting together to make a coherent
explanatory approach to phonotactic universals, this paper is intended as a brief introduction to the nature of that approach.

I. Recent work of first language acquisition

Table I. summarizes the essential points about early child phonology that are needed for our purposes. (The arguments are presented
in much more detail in Menn 1978, 1979, 1982, and in press; see also
Hastings 1981.)

A fascinating aspect of early child phonology - say the first
six months to a year of speech - is the strictness of the output
constraints that the child's words obey, and the variety of rules
which conspire to maintain those constraints. ('Rule' here is used
in the sense of a regular correspondence between adult and child
form. It is not strictly correct to speak of c child 'using' such
surface-to-surface rules, since the child's underlying form may
differ from the presented adult surface form, which is itself an
abstraction. (See Menn opp. cit., Maxwell 1984.) One of the most
common constraints is a requirement of CV alternation, which appears
to correspond to a universal tendency among adult languages and which
is therefore frequently discussed. Perhaps almost as common, however, is a requirement of consonant harmony, which surprises a good

241

Table I. Output constraints in early child phonology

1. Consonant Harmony:

A. 'deliberate', via

i)   assimilation rules: 'duck'→[gʌk]
ii) deletion/weakening rules used only in the disharmonic
     environment: 'bread'→[bɛ], but→read' [id]

B. 'accidental', via the operation of general rules"

i)   substitution: /k/→[t]: 'duck'→[dʌt], 'cake'→[teit]
ii) final deletion: 'duck'→[dʌ],   'cake'→[kej]

2. Syllable structure constraints:

A. no consonant clusters

clusters broken by epenthesis: 'blue'→[bəlu]
clusters reduced by deletion: 'blue'→[bu]
clusters coalesced: 'spot'→[fat]

B. position constraints

sibilants restricted to final position: 'snow'→nos
rigid non-harmonic consonant frame (Macken): 'sopa'→[pwɪtta]

(Metathesis examples chosen here as being most spectacular; in
all cases, a variety of deletion and substitution, selection and
avoidance strategies conspire to maintain these constraints.)

3. Final consonant voicing strategy (Fey & Gandour)

'drop'→[dapʰ]   'eat'→[itʰ]   'cook'→[gukʰ]
'stub'→[dabm̩]   'had'→[bædn̩]   'big'→[bagŋ̩]

many people (even though Jakobson mentioned it), since it is a very
minor phonotactic constraint cross-linguistically; for a survey and
comparison of the acquisition and cross-language data, see Vihman
(1978). The first part of Table 1. gives some indication of how
consonant harmony looks in early child speech; place-of-articulation
harmony may be the most frequently attested variety, but nasal harmony is also found, and some other manner harmonies seem to turn up
occasionally. It is important to note at the outset that these constraints are only common; in child phonology, as far as I know, all
statements worth making are probabilistic, including Jakobson's implications. That fact is of course one of the major forces behind
the last decade's revolution in child phonology - it has been necessary to abandon determinism for a discovery model (see Hastings 1981,
Macken and Ferguson 1983, Menn opp. cit).

The conspiratorial (Kisseberth 1970) quality of child-phonology rules is clear; one finds an absence of non-harmonic forms like 'duck' across many children, but there may be a number of types of rules that produce this end result. All of these rule types, assimilation, deletion, and substitution, may co-occur in one child.

Assimilation, when one finds it, gives direct evidence that disharmonic sequences are really being removed (representation of two-syllable words by reduplication of the stressed syllable, often thought of as the most characteristic early child output, as in /bebe/ for biscuit, is of course a very strong form of assimilation). Across children, however, there are very few disharmonic sequences even when there is no rule of assimilation or reduplication; as mentioned, some children weaken one of the offending consonants, for example to /ʔ/, which seems immune to harmony constraints. Other children have general substitutions that happen to obviate a large number of potential disharmonies – for example, a child who uses t̲ for k̲ will automatically produce a harmonic output of /dʌt/ for duck. And a good many children drop final consonants in general, so that they don't run into the problem at all, at least within the syllable.

Consonant cluster simplification is frequently accomplished by deletion, coalescence, or epenthesis, as illustrated in the second part of Table 1. These act directly to break up the structure, but again, a given child may have some other rules that act indirectly to produce the same end result – for example, a rule deleting all continuants.

We also find a number of children avoiding attempts at words which are disharmonic; avoidance of words containing consonant clusters is not yet clearly attested, but it may well occur.

Understanding child phonotaxis becomes even more challenging when we consider the child who has syllable-position constraints – for example, one who permits fricatives only in word-final position, or even one whose only word-final obstruent is a fricative, as was the case for Natalie Waterson's subject 'P.' (Waterson 1971). He had a canonical form /(C) Vš/, and used it to render 'dish', 'fetch', and 'vest', for example. This kind of fricative-final canonical form has been attested for at least four diary-studies children in the literature, including Hildegard Leopold, although they all had different ways of mapping adult words into it; some of those children, incidentally, broke up s̲+consonant clusters, as in, say, 'snow' by putting the /s/ at the end of the word; reduced to a rule, this appears a metathesis, yielding /nos/.

Equally amazing is Macken's Spanish-learning subject Si. (Macken 1979), who could break consonant harmony if and only if the first consonant or glide of her output word was labial (it could be p, b, w, or m) and the second was dental. Given an adult model which contained a labial and a dental stop, Si. would rearrange it in whatever way was necessary to get it into her pattern. For example, 'Fernando' became /wanno/ or /nanno/, 'manzana' became /mana/, and 'sopa' (with the use of /t/ for /s/), became /pwætta/!

At this point, the statement that children simplify adult words demands close reexamination. If we hold fast to the idea that children's output differs from adults' when it is simpler for them to do it their way, we must abandon the idea that one can know a priori what is going to be simpler than what else. Why should one child find /Iš/ the easy way to say 'fish' and another child find /fᴛ/ the easy way? Why didn't Macken's subject say 'topa' for 'sopa', like the standard child in any textbook? Where do these crazy output constraints come from?

The short answer is: children learn to say sequences of segments, not just segments. In general, however, that statement is too restrictive. Descriptively, what they learn are canonical forms; ways to say small sets of phonotactically similar words (Ingram 1974). In psycholinguistic terms, a child with consonant harmony can be seen as having learned an articulatory program of opening and closing her mouth that allows here to specify two things; the vowel and one point of oral closure. (A child at this stage of development is fairly likely to lack real control over obstruent voicing – the initial stops will probably be short-lag, voiced to the English-speaker's ear, and final stops will probably be automatically devoiced. In other words, voicing will not be a parameter chosen by the child; it will be an automatic consequence of the child's inability to produce initial unvoiced stops and final voiced stops).

With the notion that a whole word is a program with a couple of settable parameters, the less plausible output constraints lose some of their weirdness. At least, given that a child has once learned how to produce a monosyllabic word with an initial consonant and a final sibilant, running off the program that produces /bɑs/ for 'bush' can very likely be adjusted to produce /nos/ for snow. The amount of information that has to be specified after calling up the program is no greater than that required to produce a harmonic CVC sequence: in either case, the program has been learned, the child specifies the vowel and the point of closure, and then runs off the program. Acquisition, in this model, becomes a matter of concatenating programs to make polysyllables and learning to set more parameters within a program.

The adult end stage envisioned by this model is not one-program-for-one-segment, although it could be. The reason for not proceeding to this logical end point is the following: it's clear that we can buy a good deal of explanation of consonant cluster constraints and other output constraints if we instead assume that a tautosyllabic consonant cluster (including word boundary) is controlled by a unitary program. The proper unit might well be larger; the demisyl-

lable is a very good candidate. Evidence for these statements comes
from two sources: second language acquisition and aphasia.

But before we go on to consider those data, let us look at the
sample in part 3 of Table I, taken from Fey and Gandour's subject
Lasan (Fey & Gandour 1982). Here's a child who invented a way to
keep voicing going through a final stop by adding a final nasal.
The result looks more exotic than adding a vowel, which children
occasionally also do in this situation. Phonetically, of course,
it's a matter of opening the velum before opening the mouth instead
of afterwards. The point is that in phonotaxis there are genuine
similarities between the child learning a first language and the
older person learning a second language which has unfamiliar phones,
clusters, and/or adjacency of phones or clusters to boundaries. A
child who has even one word has a set of things she can say and a
larger set of things she wants to say; she is in a position of ten-
sion between what she has and what she needs. Like all of us whose
reach exceeds grasp, she makes use - sometimes rather unexpected use
- of what she's got, and innovates with varying degrees of success
and consistency in the attempt to fill the gap.

So do the second-language learners reported by Eckman (1982), by
Greenberg (1981), and by several contributors to this volume. The
subjects cited in Table II are like Lasan and other young children
in that they have not yet (in general) learned to keep vocal-cord
vibration going through final oral closure. They may devoice, as in
the Spanish example from Eckman (1982), or they may add a final
vowel, as in his Mandarin example, effectively bringing the original
final consonant into a syllable-initial position, in which their
native language does have a voicing contrast. Eckman has reported
that variability is very common - note that two forms are given for
the Mandarin speaker's attempt's at 'tag'; this is just what we find
in L1-learning children during transition.

The similarity of phonotactic restrictions and the applicability
of the child-phonology-based psycholinguistic model to L2 learning
is strongly reinforced by Greenberg's (1981) findings. She showed,
using a controlled comparison of Greek and Turkish learners of English, that position with respect to word boundary is an essential
aspect of a cluster. Greek forbids or restricts word-final clusters
and Turkish excludes word-initial ones except in loan-words, although
just in terms of segment identity, both languages actually allow
many of the clusters demanded by English (indeed, Greek has some
word-initially clusters, such as #ps, that English doesn't). In the
interlanguage (the intermediate stage of acquiring the second language), the Greek-born students had considerable difficulties with
final clusters, while the Turkish speakers had difficulties with initial clusters, clusters which they could say when they appeared on
the same end of a word as permitted by their native language. This
strongly supports the applicability of the acquisition-based psycho-

Table II.   Second language acquisition

1. Final voicing:

   Spanish: devoicing

       'rob'   [rap]   'robber'   [rabər]

   Mandarin: vowel addition

       'tag'   [tægə], [tæg]

2. Cluster problems:

   Japanese: vowel insertion

       'street [stərit]   'treat'   [tərit]   'try'   [traj]

   Spanish: vowel prothesis

       'splash' [ɛsplæš]   'steam' [stim], [ɛstim]

linguistic model for phonotactic constraints to adult second-language
learning.

Eckman also examined attempts at consonant clusters in his Spa-
nish-speaking/English-learning subjects. According to the psycholinguistic model that was derived from L1 acquisition, and reinforced
by the L2 findings of Greenberg 1981, the production of a consonant
cluster in a specified position (initial, final) required an articulatory program that is specifically designed to produce that cluster
in that position. Spanish speakers have articulatory programs for
word-medial sC and sC+Liquid clusters, but not for those clusters in
word-initial and word-final position. From looking at Eckman's error
data, which show vowel prothesis (very much in line with informal
experience), it indeed appears that those programs either include a
preceding vowel or can only be accessed after a vowel. That limitation is appropriate for Spanish phonotaxis, but not for English. A
similar sequential programming explanation extends to other cases of
exceptionless surface phonotactic constraints, such as those discussed by Fellbaum (this volume).

The one phenomenon (if it proves to be real) that would need
more explaining would be the case in which the target language demands a phonotactic configuration that is found in L1, but the learner fails to product it even so. The possibility of such an event
is suggested by the claim that certain trade languages show phonotactic restrictions more severe than either 'parent' language. A programming model can also be made to handle that, but to do so requires
a considerable psycholinguistic elaboration of the notion of phonotactic markedness. Such an elaboration will be sketched with motiva-

tion from the aphasia data, which we will turn to shortly. Before
that, however, the relation of perception to phonotaxis must be taken
up.

II. Articulatory 'free rides'?

Up to this point, we have treated output constraints in adult
language as if they were entirely due to the speaker's articulatory
inexperience. And that is clearly not the case. An English speaker
using babytalk or indulging in onomatopoeia can say 'thwee' for
'three' or 'thwip' like Spiderman throwing out his web. More generally, it seems to be the case that an English speaker can put a /w/
wherever she has provocalic /r/. We must conclude that some things
which one has learned to say because English demands them apparently
confer the ability to say other things that English does not demand;
experiments with artificial L2 learning are urgently needed to find
out just what those 'conferrings' are. For example, can naive
English speakers use /l/ in novel clusters that parallel 'stweet'
and 'thwee'? It shouldn't be too hard to define adequate experimental criteria and find out.

The idea that one can indeed pronounce some non-occurring consonant clusters goes beyond the SPE-type distinction between essenti-
ally-absent 'bnik' and accidentally-absent 'blik'; what shall be
done with 'bwik'? The only statement on this in formal theory, to
my present knowledge, is by Clements and Keyser (1981 and forthcoming). Their way of capturing English phonotactics uses positive
and negative output constraints. The positive constraints construct
clusters of s+consonant, consonant + liquid, and the overlay of one
of these on the other (i.e. the s+consonant+liquid clusters). These
positive constraints are interpretable as templates - essentially,
that is, as schematic production programs; Clements &Keyser in fact
propose that English speakers cannot say any cluster that is not
generated by their positive constraints. The negative constraints
then desribe and rule out particular instantiations of those templates that English happens not to demand, like 'thw' and 'bw'.
Clements and Keyser further suggest that an English speaker is able
to say all the clusters excluded by the negative constraints if required to do so in handling e.g. a loan word like 'bwana'. Experiments like the one suggested above are precisely what is needed to
test this assertion.

III. Perceptual disfavoring

If there are phonotactic gaps that are not articulatory - if an
English speaker can say 'thwip', 'svelte,' and 'vroom' - why are
such clusters absent or marginal in English? Kawasaki (1982) has
investigated the possibility of a perceptual explanation. Perhaps
sequences which are disfavored across the languages of the world are

Table III.   Perceptual motivation for output constraints

(Kawasaki 1982)

1. Sequence-internal modulation relatively small

$$b + \begin{Bmatrix} \text{liquid} \\ \text{or } w \end{Bmatrix} < g + \begin{Bmatrix} \text{liquid} \\ \text{or } w \end{Bmatrix} < d + \begin{Bmatrix} \text{liquid} \\ \text{or } w \end{Bmatrix}$$

d + j < g + j, b + j

2. Cross-sequence similarities relatively large

d + l similar to d + w, g + l

b + w similar to d + w; both similar to bu

b + j similar to d + j;

b + j similar to di and du

d + j similar to bi

hard to analyze perceptually, hard to segment and/or to keep distinct
from other sequences. (Incidentally, she begins this monograph, her
dissertation, with an excellent cross-linguistic study of monosyllabic phonotactic, and has a good list of disfavored and excluded #CV
sequences, which get left out of many phonotactic accounts.)

Kawasaki set up one acoustic measure to look at segmentability of
a C+Glide+Vowel sequence and a related one to study the similarity
between such sequences and between them and CV sequences: she took
successive slices of the formant structure (10 msec apart) beginning
with the release and looked at the sum of differences between successive formant configurations for a certain number of slices. She was
able to show that many disfavored sequences which do not appear to
pose articulatory problems are relatively hard to deal with acousti-
cally, although some languages nevertheless utilize them.

IV.   Evidence from aphasic naming errors

Some samples of aphasic naming errors are presented in Table IV.

At the Aphasia Research Center we have analyzed a fair-sized
corpus of picture-naming errors from twenty-odd patients of various
diagnostic groups, from which these examples are taken. Many of the
phonological errors result in words other than the target word; these
are phonologically more distant from their targets, on the average,
than those which result in non-words, but there is no semantic link
between the target and these outputs, so they still qualify as phonological errors. (Examples from the table include 'sphynx'/'fink',
'scroll'/'troll'.) As Blumstein (1978) has emphasized, aphasic errors are indeed matters of probability. About half of the clearly-

Table IV. Aphasic Data

1. Cluster errors. About 50% are reductions; the rest change one or both elements, sometimes producing a single related or unrelated segment. Epenthesis occurs, but is rare.

Cluster	Reduction		Substitution		Other
#pr	protractor [p'taeĕř]		pretzel	[frɛtsḷ] [krɛθl] [snitsḷ]	
#br	broom	[rul]	[skrum]		
#kr	cradle	[kejdḷ]	[brejbḷ]		
#skr	scroll	[skol] [srol] [rol]			[trol]
#sf	sphynx	[sɪnč] [fɪŋk]	[spɪŋks] [smɪŋk]		[mɪŋks] [sʌfɪŋks]

Initial singletons > clusters - not common, but appears comparable to consonant prothesis

'bed' > [brɛd]   'camel' > [trꜳempḷ]

'whistle' > [trisḷ]   'bench' > [brɛnč]

More than a cluster involved?

gl 'igloo' [ɪlgu], [ɪnglu]

sk 'escalator' [ɛksl+kejtř]

2. Sequence of attempts (Kohn & Menn 1982, Kohn 1983)

A. Succession of attempts at 'snail'

[stejl], [skejl], [stejl], [sk-], snail.

B. Succession of attempts at 'asparagus'

[ɪnspæerədə], [ɪnspæes], [ɪnspræ], [ɪnspræɛʌkʌ], [ɪnspæe],
I - N, [sprꜳektɪk], [spꜳetꜳeksə], [ɪnspæerʌsʌ], [spæer-ra],
[spꜳeratɪn], got half of it, [spæerəgʌts], [spæerəgəs],
[spꜳerəgɪts], close, [spæerəgəs], [spꜳerəgə], [spargəs],
[spəgꜳerə], [spəgꜳerəs], now I'm getting worse.

C. Succession of attempts at 'igloo'

[aj-], [ajk-], [ajgpl], [ajgp-], [ajglu],[ej-], [iglu],
[ajglu], [ɪnglu], [glu], [o], [ajglu], [ɪnglu], [li-],
[gli-], [ajglu], igloo, [iglu], igloo.

D. Succession of attempts at 'accordion'

[kɔrdʌn], [kɔrdʌnʌn], [ʌkɔrdɪdʌn], [ʌkɔrdɛ'ʌn], [ʌkɔrdɪgʌn]

pronounced consonant cluster errors that we recorded were indeed simplifications of the cluster, but within those there was much variation as to which segment was omitted - for 'scroll' we found /skol/, /rol/, and even the rare /srol/ (almost no aphasic errors - less than 1% - violate the phonotactic constraints of the language in question). The other half of the cluster errors were substitu-

tions of other clusters - 'fretzel' for pretzel, or even further
afield: 'krethel', 'snitzel'. Some errors make clusters longer - the
table shows items like 'scroom' for 'broom', and there is a scatter
of cases in which a consonant is inserted to create a cluster where
the target word had none: 'bed' becoming 'bread', for example. At
only about the same frequency, we find an error that we had predicted
would be frequent, prothesis of a consonant before an initial vowel,
as in 'coctopus' for 'octopus'. Aphasics - at least those with articulation problems - indeed have trouble beginning words with initial unstressed vowels (Gleason et al. 1975), but there appear to
be no problems with vowel-initial words per se.

Consider now the errors 'inglu' for 'igloo', 'exlicator' for
'escalator'. If these are non-random errors, they strongly suggest
that the units involved are much more than the consonant cluster
itself. 'Incl-' and 'ex-' are very common word-initial sequences in
English as compared to 'igl-' and 'esc-' respectively. If the programming model of articulatory output which we have been developing
is going to account for such errors, one type of programming unit
will probably need to be vowel plus all following consonants, i.e.
the closed demisyllable. The major 'seams' where the unit programs
are concatenated should be within the vowel and - if the model is
hierarchical, as it probably has to be the minor seams will be at
points of syllable division within the medial cluster. (The major-
seams- within-the-vowel idea was suggested for the solution of
pragmatic speech synthesis problems by Fujimura and Lovins 1978).

Looking at the long sequences of attempts produced by certain
aphasic patients gives further encouragement to this specification
for the programming model. The attempts at 'asparagus' and 'igloo'
in Table IV show both the 'in' onset for asparagus and the 'ing'
onset for igloo; these errors recurred across patients. (Spelling,
incidentally, probably also influences the direction of individual
errors - that may be the reason for the /aj/ = 'I' onset which recurs
in the 'igloo' attempts. There are other patients who keep trying
to end the word with the sound 'o' - /iglo/, for example.)

The attempts at 'accordion', Table IV, 2.D, present another error
that is very important to get into a production model. This patient
seems to need to fill the hiatus between /i/ and /ʌ/, and the various
consonants in the word appear to get into that spot (one can debate
as to the origin of the /g/. The one production, /ak rd' n/, that
doesn't have an intruding consonant was a very effortful and un-
English affair indeed. Here is where the preference for CV alternation really does show up; and that's what it seems to be, rather than
a preference for consonantal word onsets.

V.   Bringing the strands together

Let's take stock. In the model for children's early output, the

production apparatus (under normal speaking condition) takes note of
the string of segments present in a word, sends for the stored programs that it has found appropriate for such strings, sets those
parameters over which control has been acquired (for example vowel
height, consonant position, and consonant nasality), and runs the
program to produce the articulation. I claim that the adult is still
using a more streamlined and flexible version of this device, one
where several sub-programs are integrated to produce a word. In
adult production, a segmental representation is retrieved from the
lexicon; then the programs are 'called' by subsequences of the segment string. (The model has to start from that segmental retrieval
stage so that it can handle slips-of-the-tongue which shuffle conso-
natal clusters, the type 'krick skwub' for 'quick scrub', Shattuck-
Hufnagel 1982.)

A schematic representation of the model that we have been trying
to work with (Kohn & Menn 1982, Kohn 1984) is given as Figure 1.
After a segmental representation is brought into working memory, it
is transferred to pre-articulatory programming. Many slips of the
tongue, especially segmental exchange errors like the one just mentioned, would take place during this transfer. In pre-articulatory
programming, the segments are assembled into the names of programs;
the named programs are then retrieved from storage, parameters (feature values, for the adult) are set within them, they are concatenated, and sent off to articulatory programming for execution. This
model is compatible with several recent proposals in the aphasia
literature (Nespoulous et al. 1982, MacNeilage 1982.)

Execution errors of the sort aphasics make are probabilistic, but
the most probable errors must be those involving slipping into a more
accessible program - likely to be one more frequently used - and
slipping into a default value - which is the CV alternation pattern.

VI.   Markedness as an intermediate variable

Finally: Where is markedness in all of this? In a lot of different places. In acquisition, the more marked segments and sequences of segments are those which tend to be harder to discover how to
say, as I have argued at length elsewhere (opp. cit.). Kawasaki's
work gives this description a new ingredient, for learning to say
involves perception in two ways: decoding the target when others
say it, and monitoring the adequacy of one's own efforts to produce
it. Markedness - or disfavoring - arising from this source has
different properties than markedness arising from articulatory
difficulty - an English-speaking adult can say 'thwee' and so can a
thwee-year-old.

I suggest that the articulatory programs are set up in such a
way that some require more information than others. Specifically,

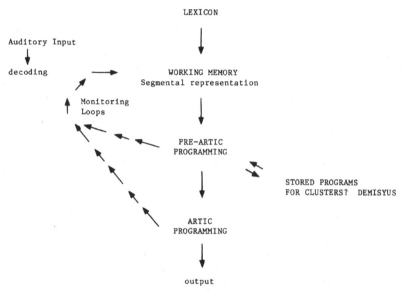

Figure 1.

the system expects CV alternation and tends to produce that as a
default when noise in the channels due to aphasia (and possibly some
other forms of stress) causes loss of information. Among the articulatory programs, furthermore, there are language-specific gradients
of acccessibility. The model that I have sketched will not actually
produce this defaulting/biased searching behavior yet, but I think
we'll have it fairly soon, and be able to begin testing it experi-
mentally.

So, again, what is markedness? All this probabilistic talk goes
against the grain for many linguists. But the child language data,
the aphasia data, the cross-language data as assembled in the UCLA
segment inventory database (e.g. Nartey 1979, Maddieson 1980), and
the L2 papers that I have cited all make it impossible to hold a
deterministic, strict-implicational view of phonotaxis.

Contrary to Jakobson (1968), a good many children do not start
to talk with the 'optimal' consonant/vowel opposition of "papa" or
"mama", or anything even close. Som children first say /mmm/, some
say 'shshshsh', some say 'hi'. Some do indeed learn to oppose /d/
and /k/ before opposing /b/ and /d/ (Menn 1976/79)./ Similarly for
aphasics: even among the very few out of the hundreds seen at our
hospital who are reduced to a single 'word', one indeed said 'bye-
bye', which is close to the theoretical 'ba ba', but another said
'muckus' and another said 'senny-fenny' over and over whenever he
tried to speak.

It is true, overall, that the Jakobson implications give the odds for languages acquisition, and it seems to be true, as Blumstein showed in her dissertation (1973), that if you set up the right features you can interpret the general tendency of aphasic errors as towards less marked feature values, and it is true that the plurality or a slight majority of aphasic phonotactic errors is towards CV-alternation (see however MacNeilage 1982, Nespoulous et al. 1983). These probabilities heavily constrain us in making psycholinguistic models of speech production. But the data demand that any markedness theory be considered as a theory of probability, as a theory of the odds, not as an absolute. That is the first major point of this survey paper.

And the second major point is that markedness is something that needs to be explained; it is not a term that explains anything, however useful it is as a description. In some areas of linguistics, hope of going beyond this level of description may be minimal at present. But in phonotaxis, so close to the periphery, so constrained by mouth and ear, I claim that it is a grave mistake in research strategy to try to collapse all the observed phenomena into neat markedness hierarchies. Some of them are discoverable effects of the devices we use to perceive and produce speech. There may also be effects due to the organization of the mind, effects we cannot explain away, biases that we can only deal with by saying 'this is more marked, this is unmarked'. But those abstract, inexplicable mental properties will only be obscured if the explicable biases are treated as though they were the same sort of phenomenon.

References

Blumstein, S. E. 1973. A Phonological Investigation of Aphasic Speech. The Hague: Mouton. Blumstein, S. (1973)

Blumstein, S. E. (1978) Segment structure and the syllable in aphasia. In Bell & Hooper.

Bell, A. and J. B. Hooper (eds.) (1978) Syllables and Segments. Amsterdam: North Holland.

Clements, G. N., and S. J. Keyser (1981) A three-tiered theory of the syllable. Occasional Paper #19, Center for Cognitive Science, MIT.

Clements, G. N., and S. J. Keyser (1983) CV phonology: a generative theory of the syllable. Cambridge, MA:MIT Press.

Eckman, F. (1984) Universals, typologies, and interlanguages. In W.E. Rutherford (ed). Language Universals and Second Language Acquisition. Amsterdam: John Benjamins Publishing Company.

Fey, M., and J. Gandour (1982) Rule discovery is early phonology acquisition. J. Child Language 9:1, 71-82.

Fellbaum, M. L. (1983) Markedness and allophonic rules. This volume.

Fujimura, O., and Loving, J. B. (1978) Syllables as concatenative phonetic units. In Bell & Hooper, pp. 107-120.

Gleason, J. B., H. Goodglass, E. Green, N. Ackerman, and M. R. Hyde
   (1975) The Retrieval of syntax in Broca's aphasia. Brain and
   Language 2:451-471.
Greenberg, C. (1981) Syllable structure in second language acquisi-
   tion. Paper read at 1981 Annual Meeting of the LSA, New York.
Hastings, A. J. (1981) Natural Equational Phonology. Doctoral
   dissertation, Indiana University.
Ingram, D. (1974) Phonological rules in young children. J. Child
   Language 1:1,49-64.
Jakobson, R. (1968) Child Language, Aphasia, and Phonological Uni-
   versals. (trans. A. Keiler) The Hague: Mouton
Kawasaki, H. (1982) An acoustical basis for universal constraints
   on sound sequences. Ph.D. dissertation, U. C. Berkeley.
Kohn, S., and L. Menn. Conduites d'approache in conduction aphasia:
   The diagnostic power of phonologically-oriented sequences. Paper
   read at the 1982 BABBLE meeting, October, Niagra Falls, Ontario.
Kohn, S. (1984) ms. The nature of the phonological disorder in
   conduction aphasia. Brain and Language 23.97-115.
Macken, M. A. (1979) Developmental reorganization of phonology: a
   hierarchy of basic units of acquisition. Lingua 49, 11-19.
Macken, M. A., and C. A. Ferguson. (in press) Cognitive aspects of
   phonological development: Model, evidence, and issues. To
   appear in K. Nelson (Ed.), Children's Language, Vol. 4 Gardner
   Press.
MacNeilage, P. (1982) Speech production mechanisms in aphasia. In
   S. Grillner et al. (Eds), Speech Motor Control. Pergamon Press.
Maddieson, I. (1980) Phonological generalizations from the UCLA
   Phonological Segment Inventory Database. In UCLA Working Papers
   in Phonetics 50. August 1980.
Maxwell, E. (1984) On determining phonological underlying represen-
   tations of children: a critique of current theory. In Elbert, M. ,
   Dinnsen, D. A., and Weismer, G. (eds.), Linguistic Theory and
   the Misarticulating Child. ASHA Monograph 22, American Speech,
   Language and Hearing Association. pp. 18-29.
Menn, L. 1976/1979 Pattern, Control, and Contrast in Beginning
   Speech. Doctoral dissertation, U. of Illinois. Circulated by
   Indiana University Linguistics Club.
Menn, L. (1978) Phonological units in beginning speech. In Bell &
   Hooper (eds)
Menn, L. (1979) Towards a psychology of phonology: Child phonology
   as a first step. In Herbert, R. (ed.) Applications of Linguis-
   tic Theory in the Human Sciences. Linguistics Department, Michi-
   gan State University.
Menn, L. (1982) Child language as a source of constraints in lin-
   guistic theory. In Obler, L. K., and Menn, L., (eds.), Exceptional Language and Linguistics. New York:Academic Press. 1982
   pp. 247-259.
Menn, L. (1983) Development of articulatory, phonetic, and
   phonological capabilities. In B. Butterworth (ed.), Language
   Production, Vol. 2, London:Academic Press. pp. 3-50.

Nartey, J. N. A. (1979) A study in phonemic universals - especially
    concerning fricatives and stops. UCLA Working Papers in Phonetics
    46. November 1979.
Nespoulous, J. L., Joanette, Y., Beland, R., Caplan, D., and Lecours,
    A. R. Is there a markedness effect in aphasic substitution
    errors? ms., Laboratorie Theophile Alajouanine, Centre de
    Recherche du Centre Hospitalier Cot-des-Neiges, Montreal.
Vihman, M. M. (1978) Consonant harmony - its scope and function in
    child language. In J. Greenberg (ed.), Universals of Human Lan-
    guage, Vol. III. Stanford: Stanford University Press.
Waterson, N. (1971) Child phonology: a prosodic view. J. of
    Linguistics 7:179-221.

Bibliographic Addendum

Note: Several studies, especially those by Kent, Lindblom, and
Ohala, in MacNeilage (1983), not available to me at the time of
writing this paper, are especially germane to the question of the
articulatory underpinnings of syllabic markedness, and are recom-
mended to all interested readers.

Kent, R. D. (1983) The segmental organization of speech. In
    MacNeilage (ed.), pp. 57-89.
Lindblom, B. (1983) Economy of speech gestures. In MacNeilage
    (ed.), pp. 217-245.
MacNeilage, P. (ed.) (1983) The Production of Speech. New York:
    Springer-Verlag.
Ohala, J. L. (1983) The origin of sound patterns in vocal tract
    constraints. In MacNeilage (ed.), pp. 189-216.

LANGUAGE ACQUISITION DATA AND THE THEORY OF MARKEDNESS:

EVIDENCE FROM SPANISH

Lawrence Solan

1.   Language Acquisition and Markedness Theory

The theory of markedness is used to account for some very different sets of facts. For example, Chomsky (1981a) defines marked structures as those not predicted by core grammar. Assuming the current
theory of core grammar to be correct, those constructions that the
theory predicts are ungrammatical, but which in fact exist, are
called marked. The presence of the NP to VP structure in English,
discussed in Chomsky and Lasnik (1977) and Chomsky (1981b) has been
explained in this way.

(1)   John believes [Bill to be the winner].

Although (1) is perfectly grammatical in English, most languages do
not permit subjects in infinitival clauses, and the current theory
of grammar, as proposed in Chomsky (1981b), does not allow such constructions in general. To account for the grammaticalness of (1),
it is proposed that verbs like believe have special properties en-
abling them to assign case to the subject of the embedded clause.
Rather than assuming the theory to be wrong, additional explanation
based on notions of markedness is proposed to account for deviant
facts.[1]   Thus, according to this view of markedness, exactly which
structures are the unmarked ones depends crucially on the theory,
rather than on a set of readily observable facts.

But facts there are. It has long and frequently been observed
that when languages differ with respect to a particular phenomenon,
one sort of language seems to be the "normal" case, while another
seems "odd" in this respect. Thus, when languages of the world generally work in a particular way, this is said to be the unmarked

257

case, a conclusion generally based on frequency. Along these lines, constructions on which grammaticality judgments may vary, or which seem to be marginal, are thought to be marked.[2]

In this paper, I would like to consider a third perspective on markedness, one which also has been recognized in the literature, at least since Jakobson (1941), and has been implicitly accepted by certain contemporary linguists (see, e.g., Chomsky (1981a)). If we regard universal grammar as the set of restrictions on hypothesis formation that limit the range of theories that children will con- sider in learning their native language, then markedness can be seen from a psychological viewpoint. In some instances, linguistic uni- versals apparently limit the possibilities to one . For example, all languages seem to have noun phrases. However, in other cases, universal grammar restricts the possible characteristics that languages may have, but leaves open more than one possibility. Here, it is said that universal grammar defines certain parameters, whose value the child will fix from his experience during the lan- guage acquisition process. Looking at linguistic theory this way, part of the theory of markedness will be the theory of which value for each parameter is the "normal" one; that is which value the child will consider first.

Looked at from this perspective, the theory of markedness becomes explanatory. For not only does the theory account for the distribu- tion of particular characteristics across the world's languages, but it also explains this distribution in terms of some events of the language acquisition process mandated by innate structures. There- fore, language acquisition research of the type discussed in this paper and in similar work is by no means intended to supersede the more conventional research strategies into the theory of markedness discussed above. To the contrary, investigations into the language acquisition process may bring corroborative, disconfirming, or per- haps entirely new evidence to bear on issues independently interes- ting to those concerned with principles of universal grammar and the theory of markedness.

Note that this view makes predictions somewhat different from the others described above. For example, it may well be that the unmarked case is by no means the most frequent. Cairns and Feinstein (1982) observe that while [a] is the only vowel in the unmarked vowel system, very few languages in the world limit them- selves to this unmarked system, with most languages having far more complicated vowel systems. Moreover, variation with respect to the value of a parameter either within a speaker or across speakers within a language may indicate that this particular parameter has no unmarked value. See Solan (1981) for discussion. The distinction between core and peripheral grammatical constructions may also play a role in this version of markedness theory, if it is indeed the case that children generally acquire core functions before peripheral

ones. In Solan (1983), where it is shown that children identify the
presence of focus before understanding its meaning in a particular case, it is argued that core functions are learned first.

2.   A Brief Look at Locality Principles

In this paper I will examine the theory and acquisition of the
locality principles that contribute to the interpretation of sentences like (2),

(2)   a.   *Paul said that Della ignored himself.

b.   *Paul asked Della to ignore himself.

and then describe some language acquisition studies, both in Spanish
and in English, that seem to shed some light on markedness and other
related aspects of linguistic theory.

First, let us consider the domain of the binding principles in
English. As (2) illustrates, the principles apply blocking binding
of reflexives across both tensed and infinitival clause boundaries.
Substituting him for himself in (2) creates grammatical sentences.
This complementarity is captured by the principles of the binding
theory listed in (3), where D is the domain in which these principles apply (see Chomsky (1981b)):

(3)   a.   An anaphor must be bound in D.   (Reflexives are
                                             anaphors.)

b.   A pronoun cannot be bound in D.[3]

The task of defining D has been a difficult problem for linguistic
theory during the past decade, with many proposals appearing in the
literature, beginning with Postal's (1971) clausemate restriction.
In English, we know that at least the data in (2) must be accounted
for. Relevant are sentences like (4), in which the reflexive or
pronoun is contained in a noun phrase.

(4)   a.   John found that stupid picture of himself.

b.   John found Bill's stupid picture of himself.

The key generalization, as pointed out in many of Chomsky's recent
writings, is that reflexives cannot be bound across a subject, whether that subject is contained in a clause or a noun phrase. In
earlier work, Chomsky (1973) had concluded that binding cannot occur
across either a tensed clause or across a subject. More recently,
he has collapsed the various environments into a single concept,
"accessible SUBJECT," which includes the agreement element of tensed
clauses, and the subject of infinitives and noun phrases.

Wayne Harbert has conducted considerable research indicating that not all languages have precisely the same binding restrictions. For example, while both of the sentences in (2) are ungrammatical in English (and Spanish), only the equivalent of (2a) is ungrammatical in many languages, including Icelandic, Hindi, Gothic and Latin. This fact makes it likely that a single notion, like "accessible SUB-JECT," is not sufficient to describe binding theory in universal grammar, since languages seem to choose among a small number of possibilities with respect to the domain D in which the locality principles apply. See, e.g., Harbert (1981).

A child learning a language will, of course, not be privy to whether he is about to learn English or Icelandic. Thus, we may say that the domain of the binding theory is a parameter, with a limited set of values that must be fixed by experience. Whether a particular value is the unmarked one is another question that must be addressed. To learn the locality principles, a child must know: (1) enough syntactic structure to identify clause and noun phrase boundaries, and make use of them; (2) the set of anaphors and pronouns in the language he is acquiring; (3) the binding principles themselves; anaphors are bound, pronouns are not bound, and (4) the domain of D of the theory for the language that the child is learning. Below I will describe some experiments that address these issues in the development of Spanish and English. The work from two completed experiments will be discussed, and further experimentation, still in progress, will be described.

3.   The Development of Locality Principles in Spanish

In Solan and Ortiz (1981), we presented 28 children ages 5 through 7, with sentences like those in (5)-(8). The children attended a bilingual school in the Boston area, and all spoke Spanish natively. They were divided into two age groups.

(5)   Juan dijo que Pedro se golpeó.
        Juan said that Pedro hit himself.

(6)   Juan dijo que Pedro lo golpeó.
        Juan said that Pedro hit him.

(7)   Juan ordenó a Pedro golpearse.
        Juan told Pedro to hit himself.

(8)   Juan ordenó a Pedro golpearlo.
        Juan told Pedro to hit him.

Children were presented with four of each type of sentence, and asked to act them out using toy animals (the noun phrases were the names of animals). Only animals with the same gender either el caballo (the horse) and el perro (the dog) or la vaca (the cow) and la oveja (the

Table 1. Percentage of Correct Responses

| Condition | | Group | | |
		Younger	Older	Total
+Tense	Refl	52	41	46
	Pron	61	73	67
-Tense	Refl	48	23	36
	Pron	45	50	47
TOTAL		51	47	49

sheep) were used in any one sentence, and the verbs were chosen from a substantial list at random. The results are presented in Table 1.

Note that the sentences with tensed clauses were acted out correctly more frequently than the sentences with infinitival clauses (56% vs. 41%). This difference is statistically significant.[4] Note also that those sentences containing pronouns were significantly easier than those containing reflexives (57% vs. 41%). Interestingly, 61% of all errors involved the child taking the first animal mentioned in the sentence and having it act upon the second noun phrase mentioned in the sentence.

Looking at the results in terms of what it takes to learn the binding principles, it appears that the children at first do have trouble with the syntactic structures of the sentences, as evidenced by the high percentage of responses in which the first animal acted on the second, regardless of the structure of the sentence. Once children stop using this word order strategy, as evidenced by the almost 50% correct response rate, they have relatively little difficulty with the binding principles.[5]

Secondly, knowing the set of anaphors and pronouns is a less trivial task than one may have expected. Children have a far easier time with pronouns than with reflexives. In fact, 14% of all errors committed can be explained by claiming that children are interpreting se as though it were lo or la. Only 7% of the total errors can be attributed to children's interpreting the pronoun as though it were a reflexive. Moreover, children have more trouble with reflexives as they get older. This can be explained by reference to the ambiguous role that the morpheme se plays in Spanish. Not only can it be a reflexive clitic pronoun, but it also serves as a dative clitic in sentences like (9), and as an impersonal marker in sentences like (10). As children get older and learn these facts, some confusion may occur.

(9)   Se lo di ayer.   To him it (I) gave yesterday.

(10)   Aquí, se habla español.   Spanish is spoken here.

The pronominal clitics <u>lo</u> and <u>la</u>, on the other hand, are unambigously
third person pronouns.

Third, children must know the binding theory itself: pronominals
are free in D, anaphors are bound in D. The fact that children get
most of the sentences correct in which they do not use the "first
noun phrase is the agent" strategy indicates that there is good
reason to believe that children have control over these principles,
probably quite early on.

Fourth and last, children must know D, the domain in which the
binding principles apply.. It is this aspect of the acquisition of
the theory that relates to markedness. Recall that languages differ
as to whether their binding principles apply across all clause boundaries, or only across tensed clause boundaries. The fact that
children acted out sentences with tensed embedded clauses correctly
more frequently indicates that children are more likely to violate
the binding principles when no tense exists. It is inviting to
conclude from this fact that languages such as Hindi and Icelandic,
in which this distinction is the correct one for adults, are unmarked with respect to this phenomenon, and that languages like
English and Spanish are the marked ones. Recall that we are approaching markedness from the viewpoint of which values children
will assign to parameters without relevant experience telling them
to do otherwise.

Despite the attractiveness of this conclusion (which I actually
made at first), at least two factors call for doubt. First, if languages like Icelandic are the unmarked ones, then it would be very
difficult for children learning English and Spanish ever to hear
primary data that would cause them to change their minds. Because
English is more restrictive than Icelandic, children learning English would have to be told by adults that binding by the matrix
subject is impossible in sentences like (7). If, on the other hand,
English and Spanish were considered unmarked with respect to this
phenomenon, then a child learning a language like Icelandic would
only have to hear sentences like (7) with <u>Juan</u> as the intended
referent to learn that he is learning a language with this parameter receiving a marked value.

Secondly, the sentences in this experiment differ in ways other
than the presence or absence of tense. For example, the position of
the clitic pronoun differs depending on whether or not it is attached to an infinitive. More important, I believe, is the fact
that understanding the sentences with infinitives requires assigning an antecedent for the missing complement subject of the embedded

clause. Generally, in Spanish as in English, it is the matrix object
that is assigned as the controller. Occasionally, however, with a
few verbs like _promise_ (_prometer_ in Spanish), it is the matrix subject that controls. Some of the sentences in this experiment contained _prometer_ as the main verb, and children frequently did not
know that the subject was the controller. This is consistent with
the early work of Carol Chomsky (1969).[6] If it is the case that
children have difficulty assigning antecedents to missing complement
subjects, then this added burden may be sufficient to cause difficulty in applying the locality principles.

In order to address this unanswered question from Solan and
Ortiz (1981), a second experiment has been conducted. Again using a
toy-moving task, children were presented with sentences like those
in (11)-(14). Again, children ranged in age from 5 to 7, and there
were 24 children who took part in the study. Verb and noun phrases
(again the names of animals) were chosen at random, each child
receiving an individually randomized set of sentences.

(11)   Juan trató de golpearse.
        Juan tried to hit himself.

(12)   Juan trató de golpearlo.
        Juan tried to hit him.

(13)   Juan dijo que se golpeó.
        Juan said that _pro_ hit himself.

(14)   Juan dijo que lo golpeó.
        Juan said that _pro_ hit him.

The questionnaries also contained sentences not relevant to this
discussion.

Spanish is a pro-drop language. That is, subject pronouns need
not be articulated. Unlike English, sentences like (15) are perfectly grammatical in Spanish.

(15)   Escribí.   Wrote (1st pers., sing., pret.).

None of the sentences in this experiment contain overt complement
subjects. Those in (11) and (12) contain PRO, the empty element
that appears only as the subject of infinitives. See Chomsky
(1982). On the other hand, the empty subjects in (13) and (14) are
these unarticulated subject pronouns, as evidenced by the fact that
the sentences would be grammatical had they contained overt pronouns
(_él_ or _ella_), or even names.

The results of this experiment are presented in Table 2.

Table 2.   Percentage of Correct Responses.

-Tense	Refl	99
	Pron	79
+Tense	Refl	82
	Pron	79
TOTAL		85

Note first that the sentence containing both an infinitival complement and a reflexive clitic pronoun, (11), is the easiest for children to interpret, children responding correctly 99% of the time.

On the other hand, when the reflexive clitic was contained in an embedded tensed clause, the children responded correctly only 82% of
the time. This difference is statistically significant. However,
the difference between infinitival and tensed clauses is neutralized
when it comes to those sentences containing pronouns, the children
responding correctly 79% of the time regardless of the type of complement. Moreover, comparing those sentences with embedded tensed
clauses, the children performed equally well regardless of whether
the sentence contained a reflexive or pronominal clitic (82% vs.
79%). Thus, the question that must be addressed is why children
have such an easy time with sentences like (11), especially in light
of the results discussed earlier in this paper.

These facts can be explained if we abandon our initial hypothesis that languages like Icelandic, in which the binding principles
distinguish between tensed and infinitival clauses, represent the
unmarked case. In this experiment, it is the infinitival complements
that are easier. Rather, it would appear that the difficulty with
infinitives in the first experiment reflects children's having difficulty assigning a complement subject. Notice that PRO in (11) and
(12) has only one possible antecedent: Juan. If children know that
the controller of a missing complement subject in an infinitival
clause must be sentence-internal, then they will have little difficulty getting sentences like (11) and (12) correct. I will return
to the difference between (11) and (12) directly.

On the other hand, the missing subject in (13) and (14), the
tensed variants, behave exactly like full pronouns. That is, the
antecedent can either be Juan, or an individual not mentioned in the
sentence at all. In a sense, then, the tensed sentences in this
experiment are like the infinitival sentences in the first experiment in that prior to applying the locality principles, a decision

must be made as to who the subject of the embedded clause is. This
may cause added difficulty, perhaps by increasing processing load,
along the line suggested by Goodluck and Tavakolian (1982). In
fact, just as children had difficulty with prometer in the first
experiment, they occasionally chose as the subject of the embedded
clause in (13) and (14) an individual not mentioned in the sentence.

If the conclusion that children have more difficulty with (13)
than with (11) because of problems assigning an antecedent for the
missing complement subject is correct, then we would also expect to
see a difference between (16), on the one hand, and (17) and (18) on
the other.

(16)   Juan dijo que Carlos se golpeó.

(17)   Juan dijo que él se golpeó.

(18)   Juan dijo que pro se golpeó.

        (Juan said that Carlos/he/pro hit himself.)

Experimentation comparing these sentence types, among others, is
currently in progress. Preliminary results indicate that this prediction will prove to be accurate.

At this point, then, we are left with the conclusion that there
is no reliable evidence indicating that children do, as a matter of
their linguistic competence, distinguish between tensed and infinitival complements in their application of the binding principles. If
it is indeed the case that they do not do so, then the facts about
language acquisition comport with expectations based on learnability:
children at first consider the more restrictive alternative, and
through experience learn languages whose principles are relaxed to
some extent. The only sure way to test this conclusion with respect
to these phenomena is to conduct similar experiments in a language
like Icelandic. In such an experiment, we might expect to find children failing to distinguish between tensed and infinitival complements, using instead the unmarked restriction, which is the correct
one in languages like English and Spanish.

A second aspect of the data in the second experiment requires
explanation. Recall from the description of the Solan and Ortiz
study that children learning Spanish had an easier time interpreting sentences with pronouns than they did interpreting sentences
with reflexives. This is clearly not the case in this study. When
the proforms are embedded in infinitival clauses, it is the reflexive that is the easier one for children to understand. On the
other hand, when the proforms are embedded in tensed clauses, no difference whatsoever exists as a function of the type of proform.
This neutralization must also be explained.

In this experiment, the sentences with reflexives have a special characteristic that relates to the task that the children were asked to perform: The acting-out of the sentence can be accomplished by manipulating only the animal specifically mentioned in the sentence. Recall that in the first experiment, all of the sentences could be interpreted without looking for outside antecedents, although choosing the correct antecedent within the sentence provided considerable difficulty. In this experiment, however, because there is one sentence-internal antecedent available, children interpreting the sentences with pronouns must look to an external antecedent, or violate the binding principles. As the difference between (11) and (12) indicates, children sometimes chose the latter course. This is not surprising, since children generally prefer sentence-internal antecedents for pronouns when a potential antecedent precedes the pronoun. See Solan (1983). The neutralization of the distinction between pronouns and reflexives in (13) and (14), the tensed variants, can be similarly explained. The added difficulty attributable to the difference in tasks may account for the fact that the sentences containing pronouns are not easier for the children to understand than those containing reflexives. It should be noted that virtually all errors in this experiment involved the child's interpreting a pronoun as though it were a reflexive, or vice versa.

4. Conclusion

Additional experimentation is currently being conducted in both English and Spanish to test the validity of some of the conclusions drawn in this paper. Preliminarily, it appears that children learning English as their first language also have an easier time with sentences containing tensed, rather than infinitival, embedded clauses. However, the pattern of responses attributable to the difference in type of proform appears to be entirely different, leading to the conclusion that language-specific factors are involved, in much the same way as described above.

Consequently, just as the presence of tense relates to aspects of universal grammar and the theory of markedness, the relative ease of acquiring the meanings of particular proforms is strictly a matter of experience, having nothing to do with linguistic universals at all. In an inquiry into some rather complex grammatical phenomena, it should not be surprising that we come upon some facts that reflect rather deep properties of the languages faculty, and others that seem to have rather superficial sources. As I hope to have shown, this sort of language acquisition work closely resembles conventional linguistic research in that hypotheses are constantly revised on the basis of new data and new insights into explaining what we seemed to have already understood.

NOTES

*This research was supported by a grant to C. Snow from the

William F. Milton Fund of Harvard University, to whom I wish to
express my gratitude. A special note of thanks goes to Reinaldo
Ortiz, who conducted the experiments. In addition, I benefited from
the comments of many participants at the Milwaukee conference, and
the final version of this paper reflects a number of changes based
on some helpful comments and suggestions.

1. This form of argument has both its strengths and its weaknesses.
On the one hand, if the theory of markedness is to play any role in
linguistic theory, it should be expected that certain marked structures, not predicted by the general theory, will exist. However,
proceeding on the assumption that the theory is correct and that the
structures generated in spite of it are marked, decreases the falsi-
fiability of any current version of theory of grammar. It thus becomes difficult in practice to distinguish between incorrect hypotheses and hypotheses correctly reflecting the existence of marked
structures. For examples of this sort of argument, see Taraldsen
(1981) and Lasnik and Freidin (1981).

2. See, e.g., Chomsky (1981a). However, as discussed in Solan
(1981), variation may be more the result of children's inability to
fix a parameter than a result of the theory of markedness. See
below for further discussion.

3. I assume for the purposes of this paper that the domains of the
two principles are the same. This is probably not accurate, as
argued by Huang (1982). However, to the extent that the domains
differ, they do so with respect to phenomena not to be discussed in
this paper. For the sake of exposition, then, it seems reasonable
to continue on the assumption that pronouns and reflexives have complementary interpretations.

4. Throughout this paper statistical significance is based on an
analysis of variance with repeated measures at the .05 level or
less.

5. These results should be compared to those of Matthei (1981), in
which it was found that children had substantial difficulty interpreting sentences with reciprocals. In sentences like (i),

    (i)   The horse said that the cows jumped over each other.

Matthei found that children frequently had the horses and cows jump-
ing over each other, as if the sentence were (ii).

    (ii)   The horses and the cows jumped over each other.

It is conceivable that the same sort of difficulties that made it
difficult for children to take advantage of the syntactic structure
in the Spanish experiment also affected the results in Matthei's

studies.   See Otsu (1981) and Solan and Ortiz (to appear) for further discussion.6.

For more recent research on the development of principles governing the interpretation of missing complement subjects by children,
see Goodluck (1981) and Hsu (1981).

References

Belletti, A., L. Brandi and L. Rizzi, eds. 1981 Theory of Marked
    ness in Generative Grammar, Scuola Normale Superiore di Pisa.
Cairns, C. and M. Feinstein 1982 "Markedness and the Theory of
    Syllable Structure," Linguistic Inquiry 13, 193-225.
Chomsky,. C. 1969 The Acquisition of Syntax in Children from 5
    to 10, MIT Press, Cambridge, Mass.
Chomsky, N. 1973 "Conditions on Transformation," in S. Anderson
    and P. Kiparsky, eds., A Festschrift for Morris Halle, Holt,
    Rinehart and Winston, New York.
Chomsky, N. 1981a "Markedness and Core Grammar," In Belletti,
    Brandi and Rizzi, eds.
Chomsky, N. 1981b. Lectures on Government and Binding, Forris
    Publications, Dordrecht.
Chomsky, N. 1982 Some Concepts and Consequences of Theory of
    Government and Binding, MIT Press, Cambridge, Mass.
Chomsky, N. and H. Lasnik 1977 "Filters and Control," Linguistic
    Inquiry 8, 425-504.
Goodluck, H. 1981. "Children's Grammar of Complement-Subject
    Interpretation," In S. Tavakolian, ed. Language Acquisition
    and Linguistic Theory, MIT Press, Cambridge, Mass.
Goodluck, H. and S. Tavakolian 1982 "Competence and Processing in
    Children's Grammar of Relative Clauses," Cognition 11, 1-27.
Harbert, W. 1981 "Should Binding Refer to SUBJECT?," paper pre-
    sented at the meeting of the North Eastern Linguistic Society,
    M.I.T.
Huang, J. 1982 Logical Relations in Chinese and the Theory of
    Grammar, unpublished doctoral dissertation, M.I.T.
Hsu, J. 1981 The Development of Structural Principles Related to
    Complement Subject Interpretation, unpublished doctoral disserta-
    tion, C.U.N.Y
Jakobson, R. 1941 Child Language, Aphasia and Phonological Uni-
    versals, Mouton, The Hague (1972 English translation).
Lasnik, H. and R. Freidin 1981 "Core Grammar, Case Theory, and
    Markedness," In Belletti, Brandi and Rizzi, eds.
Matthei, E. 1981 "Children's Interpretations of Sentences Con-
    taining Reciprocals," In S. Tavakolian, ed., Language Acquisi-
    tion and Linguistic Theory, M.I.T. Press, Cambridge, Mass.
Otsu, Y. 1981 Universal Grammar and Syntactic Development in
    Children: Toward a Theory of Syntactic Development, unpublished
    doctoral dissertation, M.I.T.

Postal, P. 1971 Cross-Over Phenomena, Holt, Rinehart and Winston,
    New York.
Solan, L. 1981 "Fixing Parameters: Language Acquisition and Lan-
    guage Variation," in J. Pustejovsky and V. Burke, eds.,
    Markedness and Learnability, University of Massachusetts
    Occasional Papers in Linguistics 6, Amherst.
Solan, L. 1983 Pronominal Reference: Child Language and the
    Theory of Grammar, Reidel, Dordrecht.
Solan, L. and R. Ortiz 1981 "The Development of Pronouns and
    Reflexives: Evidence from Spanish," Paper presented at the
    Boston University Language Development Conference.
Taraldsen, K. 1981 "The Theoretical Interpretation of a Class
    of Marked Extractions," in Belletti, Brandi and Rizzi, eds.

THE MARKEDNESS DIFFERENTIAL HYPOTHESIS:

IMPLICATIONS FOR VIETNAMESE SPEAKERS OF ENGLISH

Bronwen Benson

University of Minnesota

Eckman (1977:315) proposes a revision of the Contrastive Analysis Hypothesis (Lado 1957) which incorporates the notion of typological markedness. His proposed revision, the Markedness Differential Hypothesis (MDH), is stated below.

Markedness Differential Hypothesis (Eckman 1977)

a)   Those areas of the target language which differ from the
     native language and are more marked than the native
     language will be difficult.

b)   The relative degree of difficulty of the areas of the
     target language which are more marked than the native
     language will corrspond to the relative degree of
     markedness.

c)   Those areas of the target language which are different from
     the native language but are not more marked than the native
     language will not be difficult.

To determine markedness relations, Eckman (320) uses the notion of typological markedness which is stated as follows:

A phenomenon A in some languages is more marked than B if
the presence of A in a language implies the presence of B;
but the presence of B does not imply the presence of A.

In this paper, I will use the MDH to make predictions about the relative difficulty for Vietnamese speakers in the pronunciation of English consonant clusters. I will then report on a test of the accuracy of the predictions.

271

The categories of English consonant clusters in this paper are limited to those categories for which there exist applicable typologically-based implicational universals. The English consonant clusters within these categories have been limited to clusters composed of consonant sounds which also occur in Central and South Vietnamese dialects. This limitation excludes from the study clusters containing the interdental fricatives θ and δ and the alveo-palatal fricative $\check{s}$. The list of English consonant clusters included in the study is given in Table 1.

In contrast to the great number and variety of consonant clusters in English, Vietnamese has no consonant clusters whatsoever (Thompson 1955). The following typologically-based implicational universals proposed by Greenberg (1965: 5) will be used to determine markedness relations. Universals #2-#7 apply to consonant clusters consisting of two consonants. The words 'initial' and 'final' in the universals refer to syllable-initial and syllable-final.

Table 1.   English consonant clusters included in the study.

SYLLABLE-INITIAL CONSONANT CLUSTERS:

Unvoiced Obstruent + Liquid

pl-/pɹ- (bilabial stop)	play/pray
fl-/fɹ- (labio-dental fricative)	fly/fry
sl- (alveolar fricative)	slow
tɹ- (alveolar stop)	try
fl-/kɹ- (velar stop)	climb/cry

Unvoiced Obstruent + Nasal

sm- (bilabial N)	smoke
sn- (alveolar N)	snow

SYLLABLE FINAL CONSONANT CLUSTERS:

CONTAINING A NASAL:

Nasal + Unvoiced Obstruent

-mp (bilabial)	lamp
-nt (alveolar)	want
-ŋk (velar)	sink

Nasal + Voiced Obstruent

-mz/-md (bilabial N + alveolar obstruent)	homes/combed[1]
-nz/-nd (alveolar N + alveolar obstruent)	runs/land
-ŋz/-ŋd (velar N + alveolar obstruent)	songs/longed

CONTAINING OBSTRUENTS ONLY:

Unvoiced Fricative + Unvoiced Stop

-ft (labio-dental fricative + alveolar stop)	left
-st (alveolar fricative + alveolar stop)	last
-sk (alveolar fricative + velar stop)	ask

Table 1 (Cont.)

Unvoiced Stop + Unvoiced Fricative

-ps (bilabial stop
+ alveolar fricative)   lips/lapse
-ts (alveolar stop
+ alveolar fricative)   lots
-ks (velar stop
+ alveolar fricative)   books, axe

Voiced Fricative + Voiced Stop

-zd (alveolar fricative
+ alveolar stop)   raised

Voiced Stop + Voiced Fricative

-bz (bilabial stop
+ alveolar fricative)   robs
-dz (alveolar stop
+ alveolar fricative)   needs
-gz (velar stop
+ alveolar fricative)   eggs

Unvoiced Stop + Stop

-pt (bilabial stop
+ alveolar stop)   hoped, apt
-kt (velar stop
+ alveolar stop)   talked, act

Voiced Stop + Stop

-bd (bilabial stop
+ alveolar stop)   robbed
-gd (velar stop
+ alveolar stop)   tagged

Unvoiced Fricative + Fricative

-fs (labio-dental fricative
+ alveolar fricative)   laughs

Voiced Fricative + Fricative

-vz (labio-dental fricative
+ alveolar fricative)   saves

1. If syllables containing sequences of n consonants are to be
    found as syllabic types, then sequences of n-1 consonants
    are also to be found in the corresponding position (prevo-
    calic or postvocalic), except that CV→V does not hold.
    ([1963] 1965)

2. In initial systems, the existence of at least one cluster
    consisting of obstruent + nasal implies the existence of at
    least one cluster consisting of obstruent + liquid.

3. In final systems, the existence of at least one sequence
    consisting of a nasal (unvoiced or voiced) followed by a
    heterorganic obstruent implies the existence of at least
    one sequence of nasal (unvoiced or voiced) followed by a
    homorganic obstruent.

4. In final systems, the existence of at least one combina-

Table 2. Markedness Relations as Derived from Universals #1 - #7

Implicational Universal Number	Marked	(- with respect to -)	Unmarked
1.	C $(n>1)$	→	C_{n-1}
2.	(initial) obstruent + nasal	→	(initial) obstruent + liquid
3.	(final) obstruent + heterorganic obstruent	→	(final) nasal + homorganic obstruent
4.	(final) sonorant + voiced obstruent	→	(final) sonorant + unvoiced obstruent
5.	(final) stop + stop	→	(final) fricative + stop
6.	(final) fricative + fricative	→	(final) fricative + stop -or- stop + fricative
7.	(final) voiced obstruent + obstruent	→	(final) unvoiced obstruent + obstruent

tion of sonant + voiced obstruent implies the existence of at least one combination of sonant + unvoiced obstruent.*

5. In final systems, the presence of at least one combination of stop + stop implies the presence of at least one combination of fricative +stop.

6. In final systems, the existence of at least one fricative + fricative combination imples the presence of at least one stop + fricative or at least one fricative +stop combination.

7. In final systems, the existence of at least one combination consisting of two voiced obstruents implies the existence of at least one combination consisting of two unvoiced obstruents.

Using the notion of typological markedness as stated on page 271 above, markedness relations can be determined from the above universals. These markedness relations and the numbers of the universals from which they are derived are given in Table 2.

The first implicational universal establishes that consonant clusters are marked with respect to single consonants. As there are no consonant clusters in Vietnamese, the whole range of English consonant clusters may then be considered marked relative to the single

*In the text, Greenberg specifies sonants as the class of semi-vowels, nasals, and liquids. Hereafter, I will use the term sonorants to refer to that class.

consonants of Vietnamese.[2] According to part (a) of the MDH, the
marked consonant clusters of English should be difficult for the
Vietnamese learner of English.   However, not all of the English consonant clusters are equally marked.   And the degrees of markedness
of the clusters must be determined in order to use part (b) of the
MDH to make predictions about the relative degree of difficulty of
the clusters.   In this case, since there are no Vietnamese clusters
with which the English consonant clusters may be compared, the implicational universals will be used to determine the relative markedness of the various English consonant clusters with respect to each
other.

The set of English consonant clusters in this paper will now be
divided into three groups: 1) syllable-initial consonant clusters
2) syllable-final clusters containing a nasal, and 3) syllable-final
clusters consisting of obstruents only.   Separate hierarchies of
markedness will be established for each of the three parts. I am
making this division because I found no pertinent implicational universals connecting initial with final clusters or connecting clusters containing a nasal with clusters containing obstruents only.
Below are the hierarchies of markedness for each of the three groups
of clusters.

HIERARCHY OF MARKEDNESS: SYLLABLE-INITIAL CLUSTERS

    pl   pɹ-   fl-   fɹ-   sl-   tɹ-   kl-   kɹ-   sm-   sn-

    Levels of Markedness

1. pl- pɹ- fl- fɹ- sl- tɹ- kl- kɹ-} obstruent + liquid

2. sm- sn- } obstruent + nasal

The clusters of the first level are the least marked clusters of
this group.   They are unmarked with respect to the clusters of level
2 in that they are of the relatively unmarked obstruent + liquid
type (Universal 2).

The clusters of level 2, sm- and sn, are marked with respect to
the clusters of level 1, as they are of the relatively marked obstruent nasal type (Universal 2).

HIERARCHY OF MARKEDNESS: SYLLABLE-FINAL CLUSTERS CONTAINING A NASAL
(see table, next page.)

The clusters of level 1, -mp, -nt and ŋk, are the least marked
of this group because their final obstruents are both unvoiced (Universal 4) and homorganic (Universal 3).

HIERARCHY OF MARKEDNESS: SYLLABLE-FINAL CLUSTERS CONTAINING A NASAL

⁻mp ⁻nt ⁻ŋk ⁻md ⁻mz ⁻nd ⁻nz ⁻ŋd ⁻ŋz

Levels of Markedness

Least
marked   1.   ⁻mp ⁻nt ⁻ŋk   nasal + <u>unvoiced homorganic obstruent</u>

⬇   2.   ⁻nd ⁻nz   nasal + <u>voiced homorganic obstruent</u>

most   3.   ⁻md ⁻mz ⁻ŋd ⁻ŋz   nasal + <u>voiced heteroganic obstruent</u>
marked

The clusters of level 2, ⁻nd and ⁻nz, are marked with respect to
the clusters of level 1 and unmarked with respect to the clusters of
level 3. Relative to the clusters of level 1, they are marked for
voicing of the final obstruent (Universal 4).

The clusters of level 3 are the most marked clusters of this
group. They are marked with respect to the clusters of level 2, for
having heterorganic final obstruents (Universal 3).

HIERARCHY OF MARKEDNESS: SYLLABLE-FINAL CLUSTERS - OBSTRUENTS ONLY

⁻ft ⁻st ⁻sk ⁻ps ⁻ts ⁻ks ⁻pt ⁻kt ⁻zd ⁻fs ⁻bz ⁻dz ⁻gz ⁻bd ⁻gd

Levels of Markedness

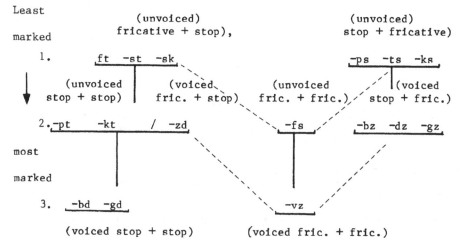

The brackets and solid lines connect groups of clusters
between which there exists a markedness relation, such that

the clusters of the lower group are marked with respect to
the clusters of the higher group.

- - - - The dotted lines connect groups of clusters for which there
exists an implicational universal, such that the existence
of the lower cluster group implies the existence of the
higher cluster group - one of two parts - but the universal
does not specify which part is implied. The dotted lines
connect the lower, marked group with each of the two parts.

The clusters of level 1 are the least marked clusters of this
group. They are unmarked with respect to the clusters of level 2
and the clusters of level 3 (Universals 5, 6, 7).

The clusters of level 2 are all marked in one respect relative
to the clusters of level 1 and unmarked relative to the clusters of
level 3. The clusters -pt and -kt, as stop + stop combinations, are
marked relative to the fricative + stop clusters of level 1 (Universal 5). The voiced cluster -zd is marked relative to the voiceless
fricative + stop clusters of level 1. The cluster -fs, as a fricative + fricative combination is marked with respect to the general
type of cluster in level 1, stop + fricative and fricative + stop
clusters (Universal 6). The voiced clusters -bz, -dz and -gz are
marked relative to their voiceless counterparts in level 1 (Universal 7).

The clusters of level 3 are the most marked of this group. The
clusters -bd and -gd are marked for voicing relative to their voiceless counterpart -pt and -kt (Universal 7); and they are marked as
stop + stop combinations relative to -zd (Universal 5). The cluster
-vz is marked for voicing relative to its voiceless counterpart -fs
(Universal 7); and, as a fricative + fricative combination, it is
marked relative to the level 2 cluster type - fricative + stop and
stop + fricative (Universal 6).

Part (b) of the MDH predicts that 'the relative degree of difficulty of the areas of the target language which are more marked than
the native language will correspond to the relative degree of markedness.' Accordingly, the relative degree of difficulty for the
Vietnamese speaker in the pronunciation of the English consonant
clusters in each of these three groups is predicted to be represented by the hierarchies of markedness. The next page lists the predicted orders of difficulty for each of the three groups.

A survey was conducted for the purpose of testing the accuracy
of the predictions made by the MDH.

The survey consists of 70 English words. These words were read
aloud by Vietnamese subjects while I transcribed their actual pronunciation of the consonant clusters. The percentages of correct pro-

nunciations were then tabulated and compared. If the predictions
hold, the order of increasing percentage of incorrect pronunciation
should correspond to the predicted order of increasing relative difficulty.

There are 35 consonant clusters represented in the survey; each
cluster occurs in two words (Appendix A).

PREDICTED ORDER OF DIFFICULTY: SYLLABLE-FINAL CLUSTERS

pl- pɹ- fl- fɹ- sl- tɹ- kl- kɹ- sm- sn-

       Levels of Markedness

increasing   1. pl- pɹ- fl- fɹ- sl- tɹ- kl- kɹ-
difficulty   ↓
           2. sm- sn-

PREDICTED ORDER OF DIFFICULTY: SYLLABLE-FINAL CLUSTERS CONTAINING A
NASAL

-mp -nt -ŋk -nd -nz -md -mz -ŋd -ŋz

      Levels of Markedness

          1. -mp -nt -ŋk
   ↓
          2. -nd -nz
          3. -md -mz -ŋd -ŋz

PREDICTED ORDER OF DIFFICULTY: SYLLABLE-FINAL CLUSTERS - TWO
OBSTRUENTS

-ft -st -sk -ps -ts -ks -pt -kt -fs -zd -bz -dz -gz
-bd -gd -vz

      Levels of Markedness

  increasing 1. -ft -st -sk -ps -ts -ks
  difficulty
      ↓   2. -pt -kt -zd -fs -bz -dz -gz
          3. -bd -gd -vz

In choosing words for inclusion in the survey, three criteria
were applied. First, all consonants and contexts of consonants,
other than the one in question, should conform to the rules of Vietnamese phonology. Second, the word should be one which the subject
is likely to be familiar with. In the survey, 66 of the 70 words
met these two criteria and four met only the second. Third, in no
pair of words containing the same target cluster should both of the
words fail to meet the first two criteria. The syllable-final cluster -sp was eliminated from the survey at this point because I was
unable to find a pair of words that met the third criterion.

These criteria were used in an effort to minimize irrelevant and
negative influences on the subjects' efforts to pronounce the words
correctly. The subjects' efforts, then, would not be hindered or
distorted by the presence of a secondary new sound or context or
unfamiliarity with the word.

In determining the order of items in the survey, my aim was to
achieve an order which would cause the least frustration to the subjects. With this in mind, I chose to put pairs of words containing
the same target cluster adjacent to each other in the survey. The
first word and the articulation of it would not be counted in the
survey results. Only the subjects' articulation of the target cluster in the second word in each pair would be counted in the results.
Then, groups of words containing the same cluster type were placed
in random order. Finally, I chose to begin the survey with a group
of the easiest clusters in hopes of easing some of the initial ten-
sion which the subjects undoubtedly would feel.

The twelve subjects who took part in the survey, ten males and
two females, were all between the ages of eighteen and twenty-nine,
and all were high-school graduates. Nine of them were enrolled in
an intensive English program in General College at the University of
Minnesota. Two more were graduates of that program. Only one subject had never been enrolled in an ESL course, and he was a very articulate, self-taught speaker of English.[3] All of the subjects
spoke the dialects of Central or South Vietnam as their primary dialect.[4]

In conducting the interviews, I attempted to control the situation in several ways which I believed would allow an accurate picture of the subjects' basic pronunciation abilities. In general,
this meant minimizing factors which would have a negative effect on
their pronunciation. All of the interviews were conducted in private. The subjects were given their copy of the survey and enough
time to read the list of words before beginning the interview. They
were then asked to read the words aloud, at their own pace. If the
subjects mispronounced the target cluster in the first of a pair of
words, I would stop them from proceeding to the second word pair,
give the correct pronunciation of the mispronounced practice word,

and ask them to try it again. I conducted the interview in this
manner so that mispronunciations which were the result only of con-
fusing orthography or lack of knowledge of the correct pronunciation
of the cluster could then be eliminated, thereby allowing a more
accurate reading of the subject's ability to produce the sounds.

Graphic representations of the survey results are presented
below.

Syllable-Initial Consonant Clusters

% Correct Articulations			Predicted Order of Difficulty				Average % Correct Artic
100	pɹ- fl- fɹ- sl- tɹ- kl- kɹ- sn-		pl- pɹ- fl- fɹ- sl- tɹ- kl- kɹ-				99
92	pl-						
83	sm-		sm- sn-				91

In the syllable-initial cluster group, the percentage range of
correct articulations is very small, only 17 percentage points.
With 100% correct articulations are seven of the eight obstruent +
liquid clusters, predicted to be the easiest clusters of this group.
The cluster pl-, also of the obstruent + liquid type, received 92%.
The average percent correct articulations of the obstruent + liquid
clusters is 99%. The results of the obstruent + nasal category,
predicted to be more difficult than the first category, show less
conformity. While sm-, at 83%, has a lower percentage of correct
articulations, sn-, at 100%, does not.

As a test of the accuracy of the predictions regarding the two
levels of consonant clusters in this group, I feel that the results
here are inconclusive, due to the fact that the range of percent
correct articulation is so small. More significant results might be
obtained in this group of syllable-initial clusters if the subjects
used had less training in English pronunciation, thereby giving a
broader range of results in this group. Regarding the gap between
the percentages of correct pronunciations of sm- and sn-, it may be
useful to note that sn- is a homorganic obstruent + nasal cluster.
In the case of syllable-final clusters, the nasal + homorganic obstruent cluster is the unmarked of the pair, and as such, is predic-

ted to be less difficult than the nasal + homorganic obstruent cluster. Although the implicational universal pertaining to such combinations applies only to syllable-final systems, Greenberg (14) states
that 'a similar statement for initial systems holds in almost all
cases'. This could explain the difference in the percentages of correct pronunciation of the clusters sm- and sn-, and would be consistent with the predictions of the MDH.

The results of the second group, syllable-final consonant clusters containing a nasal, are given in graphic form, with the predicted order of difficulty, below. These results show a much wider percentage range, 75 percentage points, and provide a better test of
the accuracy of the predictions. In this group, the results of the
survey correspond well with the predictions made by the MDH. The
nasal + unvoiced homorganic obstruent clusters, -mp, -nt and ŋk,
predicted to be easiest, are in fact the ones with the highest percentages of correct articulations - 92% for all three clusters. The
percentages of correct pronunciations of the second cluster type,
nasal + voiced homorganic obstruent clusters, -nd and nz, as predicted, are lower than the least marked cluster type and higher than
the most marked cluster type. The average percentage of correct
pronuncations for this cluster type is 71%, as compared with an average of 92% for the least marked cluster and 33% for the most marked
clusters. The third and most highly marked cluster type in this
group is nasal followed by voiced heterorganic obstruent. This type
is therefore predicted to be the most difficult. Although the percentage range of this cluster type is broad, from 17% to 50%, all of
the individual values are lower than the values of any of the less
marked clusters; and the average of the percent correct pronunciations of this type is 33%, significantly lower than the average
value of the two less marked cluster types.

In the third group, syllable-final clusters composed of obstruents only, the results again correlate well with the predictions
made by the MDH. This group has the greatest percentage range, 92
points, from 8% at the lower end up to 100%. The results of the
first custer type, voicelsss fricative + stop and stop + fricative,
show percentages of correct pronunciations ranging from 70 to 100%.
This cluster type was predicted to be easiest, and the results bear
out the predictions; with an average percent correct pronunciation
of 86%, it has the highest average of the three levels of clusters
in this group.

The clusters of the second markedness level are voiceless stop +
stop clusters pt- and -kt, voiceless fricative + fricative cluster
-fs, and voiced fricative + stop and stop + fricative clusters -zd,

SYLLABLE-FINAL CLUSTERS CONTAINING A NASAL

% Correct Articulations		Predicted Order of Difficulty	Average % Correct Artic
100			
92	⁻mp ⁻nt ⁻ŋk	⁻mp ⁻nt ⁻ŋk	92
83			
75	⁻nz		
67	⁻nd	⁻nd ⁻nz	71
58			
50	⁻ŋd		
42		⁻ŋd ⁻ŋz ⁻md ⁻mz	33
33	⁻mz ⁻ŋz		
25			
17	⁻md		
8			
0			

⁻bz, ⁻dz and ⁻gz. The MDH predicts that these clusters will be more difficult than the clusters of the first level, which are less marked, and less difficult than the clusters of the third level, which are more marked. The percentages of correct articulations of

Syllable-Final Clusters Composed of Obstruents Only

% Correct Articulations		Predicted Order of Difficulty	Average % Correct Artic
100	⁻ps		
92	⁻ft ⁻ks		
83	⁻ts	⁻ft ⁻st ⁻sk ⁻ps ⁻ts ⁻ks	86
75	⁻st ⁻sk		

% Correct Articulations	Predicted Order of Difficulty	Average % Correct Artic
67		
58		
50		
42	-bz -pt	
33	-zd -kt -fs	-pt -kt -zd -fs -bz 33
		-dz -gz
25	-gz -dz -vz -gd	
17		
8	-bd	-bd -gd -vz 19
0		

the clusters in this second level range from 25% to 42%, with an average of 33%. Corresponding to the predictions, this average of percent correct pronunciations is less than the average value for the less marked clusters and greater than the average value for the more marked clusters, 86% and 19% respectively.

The third level of clusters includes the voiced fricative + fricative cluster -vz and the voiced stop + stop clusters -bd and -gd. The clusters of this level are predicted to be the most difficult in this group. Percentages of correct pronunciations for this level range from 8% to 25% and, in keeping with the predictions, the average % correct pronunciations is 19%, the lowest of the three average values.

These data are arranged in another form (Table 3). This table gives the percentages of correct pronunciations of the clusters in the three groups, with the markedness level of each cluster in superscript. The greater the number of the markedness level, the greater the predicted difficulty.

Further support for the validity of Eckman's hypothesis can be found in an analysis of the types of pronunciation errors made by the Vietnamese subjects in the survey interviews. These errors are categorized in Table 4 according to the manner in which the original clusters were altered. Fully 97% of the errors made resulted in sequences which were less marked than the ones the subjects were at-

Table 3.

% Correct Articulations	Syllable-initial Consonant Clusters	Syllable-final Consonant Clusters Containing a Nasal	Syllable-final Consonant Clusters Obstruents Only
100	pɹ$^{-1}$ fl^{-1} fɹ$^{-1}$ sl^{-1} tɹ$^{-1}$ kl^{-1} kɹ$^{-1}$ sn^{-2}		
92	pl^{-1}	−mp^{1} −nt^{1} −ŋk^{1}	−ps^{1}
83	sm^{-2}		−ft^{1} −ks^{1}
75		−nz^{2}	−ts^{1}
67		−nd^{2}	−st^{1} −sk^{1}
58			
50		−nd^{3}	
42			−bz^{2} −pt^{2}
33		−mz^{3} −ŋz^{3}	−zd^{2} −kt^{2} −fs^{2}
25			−gz^{2} −dz^{2} −vz^{3} −gd^{3}
17		−md^{3}	
8			−bd^{3}
0			

In the table above, the markedness levels of the consonant clusters appear in superscript. The greater the number of the markedness level, the greater the predicted difficulty.

tempting to pronounce. Only one error, 0.7%, resulted in a more marked sequence, and 2.3% resulted in sequences which were equally marked.

In this study, Eckman's Markedness Differential Hypothesis was used to make predictions about the relative difficulty for Vietnamese speakers in the pronunciation of three groups of English consonant clusters. Markedness relations between the various types of clusters within each group were derived from a set of implicational universals proposed by Greenberg (1965). A survey was conducted for the purpose of testing the predictions. The survey results gave percentages of correct pronunciation of the consonant clusters, and these results, on the whole, bore out the accuracy of the predictions made by the MDH. Furthermore, an analysis of the subjects' incorrect articulations showed an overwhelming tendency toward the production of less marked sequences, providing additional support for the validity of Eckman's hypothesis.

Finally, in contrast to the Contrastive Analysis Hypothesis (Lado 1957), the predictions made by the MDH were not only basically accurate but also refined. The Contrastive Analysis Hypothesis claims that the elements of the target language which are similar to the native language will be easy, while the elements of the target language which are different than the native language will be difficult. Had the Contrastive Analysis Hypothesis alone been used in this study to make predictions, it would have predicted only that the English

Table 4.

% of total errors	Number of errors	
48.6	71	1. **Elimination of the tautosyllabic cluster structure:**

a) by insertion of a schwa between the constituents of a cluster, forming a new syllable, (cc→cəc), e.g., sm→sə-m

b) by insertion of a schwa after the final obstruent in a syllable-final cluster, resulting in a new final syllable -cv, (cc#→c-cə#), e.g., -kt→-k-tə

c) by elimination of one of the consonants in a cluster, e.g., -dz→-z

| 30.1 | 44 | 2. **Devoicing** |

a) of the entire voiced final obstruent cluster, e.g., -bz→-ps

b) of the final voiced obstruent in a nasal + obstruent cluster, e.g., -nd→-nt

| 12.3 | 18 | 3. **Elimination of the tautosyllabic cluster structure accompanied by devoicing of one or both constituents:** |

C = voiceless→c = voiced   e.g.,

c-C#	-gz →	-g-s
cəC#	-dz →	-dəs
c-Cə#	-gd →	-g-tə
C #	-gd →	-k
cc#→ C-c#	-hz →	-p-z
Cəc#	-bd →	-p-t

% of total errors	Number of errors	
5.7	9	4. **Substitution of one of the cluster constituents with a different consonant, resulting in a less marked cluster:**

a) (6)-fs → -ps (fricative + fricative → stop + fricative)

b) (1)-gd → -zd (stop + stop → fricative + stop)

c) (1)-ŋd → -nd (nasal + heterorganic stop → nasal + homorganic stop)

d) (1)-pt → -ps (stop + stop → stop + fricative)

| 2.7 | 4 | 5. **Substitution of one of the cluster constituents with a different consonant, or reversal of constituent order, resulting in a cluster which is not less marked:** |

Equally Marked:

a) (1)-ts → -st (stop + fricative → fricative + stop)

b) (1)-sk → -ks (fricative + stop → stop + fricative)

c) (1)pl- → fl- (stop + liquid → fricative + liquid)

More Marked:

a) (1)-ps → fs (stop + fricative→ fricative + fricative)

consonant clusters would be difficult for Vietnamese speakers to
acquire. The MDH, on the other hand, went beyond this prediction to
give a refined set of predictions regarding levels of difficulty
among the various types of English consonant clusters.

Notes

I would like to thank Jeanette Gundel, Kathleen Houlihan, Nancy
Stenson and Gerald Sanders for reading and commenting on earlier
verions of this paper.

1. N + /b/ and N + /g/ are systematically excluded syllable-finally
    in English.

2. The universals are context-specific; that is, they apply to
    syllable-initial and syllable-final consonant clusters indepen-
    dently. Consequently, the markedness relations established
    between the consonant clusters of English and the single conso-
    nant clusters of Vietnamese are these: the syllable-initial
    consonant clusters of English are marked with respect to the
    syllable-initial single consonants of Vietnamese and the syl-
    lable-final consonant clusters of English are marked with
    respect to the syllable-final single consonants of Vietnamese.

3. Although the subjects' backgrounds in English language instruc-
    tion varied greatly, all had attained roughly the same level of
    proficiency in English. Three subjects had attended high school
    in the U.S. for at least three years and received diplomas from
    those schools. Six other subjects had studied English in Viet-
    namese secondary schools for lengths of time ranging from one to
    five years. All of these nine subjects had been placed by their
    scores (on an English language proficienty test, the Michigan
    Test) in the General College Commanding English program. At the
    time of the interview, all nine had completed one quarter of in-
    tensive English instruction in the program, including one course
    in English pronunciation. Of the remaining subjects, two were
    graduates of that program, which was the only formal training in
    English they had had. One subject was a self-taught speaker of
    English. Only two subjects had studied or spoke a third lan-
    guage, and those two subjects had studied French for several
    years in Vietnamese secondary schools.

4. 1 found no differences between the dialects of Central and South
    Vietnam which were pertinent to this study.

5. Before conducting the interview, I made a judgement which had a
    bearing on the survey results. This judgement involved the
    'correctness' of unaspirated articulations of pl-, pɹ- , kl- and
    kɹ -. In standard English, the aspiration of the initial stop

in these clusters is manifested as voicelessness in the following liquid. When the Vietnamese articulate these cluters, they
generally do not aspirate the stop or devoice the liquid.
Although their articulations of these clusters do not sound like
the standard English version, they are easily recognizable as
pl-, pɹ- , kl-, and kɹ -. Furthermore, the implicational
universal in this case does not specify voicing or voicelessness
of the liquid as a condition for their unmarked status. For
these reasons, I judged that both voiced and unvoiced
articulations of the liquid would count as correct.

6.   I have also analyzed the results of the survey for possible correlation between number or type of error and the morphological
structure of the items in the survey.

All of the words in the survey which contained syllable-initial
clusters were monomorphemic.

In the category of syllable-final clusters containing a nasal,
five of the six words with nasal + voiceless obstruent clusters
were monomorphemic and one was polymorphemic. Results were the
same for both the polymorphemic and monomorphemic words of the
pair.

Of necessity, the words containing the clusters -md and -nd
(nasal + heterorganic obstruent) were polymorphemic. The words
containing the cluster -nd were monomorphemic. While the number
of errors produced in the articulations of these clusters dif-
fered, as predicted, the types of errors involved either devoicing or elimination of the final obstruent. Only one error, in
the articulation of -md, resulted in the sequence Nəd, which may
be attributable to the polymorphemic status of the word in which
the cluster appeared. Two errors in the articulation of -nd
resulted in sequences which altered the heterorganic nature of
the combinations (-ŋk, --nd); two more involved insertion of
a schwa after the final obstruent in -ŋd. I do not believe that
these errors have any relation to the polymorphemic status of
the words.

Of the words containing the syllable-final clusters composed of
obstruents only, all but six were polymorphemic.

In the least marked category of this group, five of the six
words containing voiceless fricative + stop clusters were
monomorphemic, while all six words containing voiceless stop +
fricative clusters were polymorphemic. As the percentages of
correct pronunciations indicate, there is no pattern which
correlates to the morphological structures of the words. Also,
the same types of errors occurred in both groups.

In the articulation of the cluster -kt in 'act' and 'talked',
more errors occurred in the monomorphemic 'act' (8) than in the
polymorphemic 'talked' (5).

In conclusion, I found no correlation between the morphological
structures of the words in the survey and the type or number of
errors produced.

References

Eckman, Fred R. 1977. Markedness and the Contrastive Analysis Hypo-
    thesis. Language Learning. 27.2 315-330.
Greenberg, Joseph H. 1965. Some generalizations concerning initial
    and final consonant sequences. Linguistics. 18. 5-32.
_____1963. Memorandum concerning language universal. Universals of
    Language. Edited by J.H. Greenberg. 2nd edition, 1966.
    Cambridge, Mass: The MIT Press. xxv. [Reprinted in 1965]
Lado, R. 1957. Linguistics Across Cultures. Ann Arbor, Michigan:
    The University of Michigan Press.

Vietnamese Reference Grammars:

Dauphin, Antoine. 1977. Cours de Vietnamien. Mayenne, France:
    Joseph Floch Maitre-Impreiteur.
Nguyen-Dang-Liem. 1970. Vietnamese Pronunciation. Honolulu:
    University of Hawaii Press.
Thompson, Laurance C. 1955. A Vietnamese Grammar. Seattle:
    University of Washington Press.

APPENDIX A

Words Used in the Survey:

I.   Words Containing Syllable-Initial Clusters

        pl- play/pray   sm- smoke/smell

        pɹ- pray/pro   sn- snow/snake

        fl- fly/flew

        fɹ- fry/free

        sl- slow/slip

        tɹ- try/true

        kl- clue/climb

        kɹ- cry/crack

II.   Words Containing Syllable-Final Clusters - Nasal + Obstruent

        -mp lamp/bump   -nz means/runs   -mz homes/times

        -nt don't/want   -nd land/wind   -md named/combed

        -ŋk sink/tank   -ŋz rings/songs

                                                   -ŋd longed/winged

III.   Words Containing Syllable-Final Clusters - Obstruent +
        Obstruent

        -st lost/missed   -pt hoped/taped   -bd robbed/fibbed

        -ft gift/left   -kt talked/act   -gd begged/tagged

        -sk ask/risk   -zd amazed/raised   -vz waves/lives

        -ps lips/hopes   -fs coughs/laughs

        -ts boats/hates   -bz fibs/robs

        -ks likes/books   -dz needs/roads

                             -gz eggs/bags

The first word of each pair is the 'practice' word.

MARKEDNESS AND ALLOPHONIC RULES

Marie L. Fellbaum

University of Minnesota

In 1977 Eckman argued that Lado's Contrastive Analysis Hypothesis (CAH) could be "maintained as a viable principle of second language acquisition" it if was revised to incorporate certain principles from universal grammar. Lado (1957) had stated that:

"...in the comparison between native and foreign language lies the key to ease or difficulty in foreign language learning...

We assume that the student who comes in contact with a foreign language will find some features of it quite easy and others extremely difficult. Those elements that are similar to his native language will be easy for him and those elements that are different will be difficult." (p. 2)

Eckman argued that the Contrastive Analysis Hypothesis (CAH) should be reformulated, because it made the wrong predictions in certain language situations. For example, the grammars of English and German differ with respect to word final voiced obstruents. In spite of this difference, German speakers have difficulty learning the word-final voiced stops of English, while the English speakers learning to produce voiceless stops in this position in German do not. Eckman pointed out that since voiced stops in word-final position are more marked than voiceless stops in this position, if "degree of difficulty" in learning to produce these "elements" in this position were defined according to the notion of "typologically marked", then the "empirically correct predictions can be made." As a theoretical model for second language acquisition, this was especially appealing because the notion of markedness provides a possible explanation which is "both independent of any given language (i.e., ...universal)" as well as being "independent of the facts surroun-

ding language acquisition" (p. 320). Eckman (p. 321) thus proposed
the following revision to the CAH, the Markedness Differential Hypothesis:

1) MARKEDNESS DIFFERENTIAL HYPOTHESIS

The areas of difficulty that a language learner will have can be
predicted on the basis of a systematic comparison of the grammars
of the native language, the target language and the markedness
relations stated in universal grammar, such that,

a) Those areas of the target language which differ from
    the native language and are more marked than the
    native language will be difficult:;

b) The relative degree of difficulty of the areas of the
    target language which are more marked than the native
    language will correspond to the relative degree of
    markedness;

c) Those areas of the target language which are different
    from the native language, but are not more marked than
    the native language will not be difficult. (p. 321)

The MDH has met with few problems in predicting the ease or
difficulty in acquiring phonemes in a given environment in a second
language. Phonemes, or sounds which contrast, are sounds which
occur in the same environment and are meaning-differentiating. Both
examples presented by Eckman, the German-English case discussed
above and one involving French and English, to be discussed later,
are cases involving a difference in the facts about phonemes and
positions of contrast between a native (NL) and target (TL) language.

However, language is composed of much more complex relations than
simply a set of sounds which contrast in a certain environment.
Sounds which occur in the same environment but are not meaning-differentiating and are predictable from context are called allophones.
These two categories, phonemes and allophones from both the native
and target language, together interact to cause interference in acquiring a second language. Linguists have classified this difference
in the ease or difficulty of learning the phonological categories and
system of a new language into three main components:

(1) competing phonemic categories of the NP and TL systems;
(2) the allophonic membership of the phonemic categories; and
(3) the distributions of the categories within their respective
     systems.   (Brière, 1966, p. 13)

Eckman addressed himself to the first and last interactions (with
respect to phonemes) when proposing the MDH. His data and discussion

was limited to phonemes and the acquisition and suppression of contrasts; therefore, it is only this relation which is exemplified with respect to markedness and universals in his article. Although no mention of allophones _per se_ is made, the MDH does state that "areas of difficulty" for language learners can be "predicted on the basis of a systematic comparison of the grammars". It can be assumed, therefore, that since allophones are members in the complex relations of a language, or the grammar, they are meant to be included in the predictions made by the MDH.

It is the interaction of phonemes and their allophones which will be examined and discussed in this paper. Several different interactions are possible, for example: phonemic in NL vs. allophonic in TL; or allophonic in NL vs. phonemic in TL. The first case, where the TL segment to be acquired is more marked than the non-alternating phonemic segment in the NL, has not proven to be problematic for the MDH. For example, in Modern Athenian Greek voiceless stops show no allophonic variation. As predicted by the MDH, Greek speakers experience difficulty in acquiring the more marked aspirated stops word-initially in languages, such as English, which have allophonic variation in this environment (Koutsoudas & Koutsoudas, 1962). We will restrict ourselves to the latter situation in this interaction: the unlearning of a marked allophone in the NL and the acquiring of a corresponding unmarked phoneme in the same environment in the TL.

Data from three language learning situations will be presented: English to Portuguese; Icelandic to English or German; and Polish to English. We will begin with a review of the basic notions from universal grammar involved in Eckman's proposed revision the CAH, with particular reference to those principles relevant to the three language learning situations listed above. We will then present data from each of the three cases, all involving a segment which functions as an allophonic variant in the NL but corresponds to a non-alternating phoneme in the TL. Finally, we will conclude with a discussion which compares this data with the data presented by Eckman.

Basic Notions of the MDH

The MDH appeals to two basic notions from universal grammar: typological markedness and implicational relationships. Eckman proposed that the notion of typological markedness corresponded to "relative degree of difficulty" in learning a second language and could be used to predict which direction of learning between two languages would be more difficult. Eckman's use of typological markedness follows the commonly accepted definition presented in (2):

(2) TYPOLOGICAL MARKEDNESS

A phenomenon A in some language is more marked than B if the presence of A in a language implies the presence of B; but the presence of B does not imply the presence of A (Eckman, p. 320.)

In order to establish the difference in difficulty between two "elements" or features of a particular area of a grammar, implicational universals are needed. Examples of various types of implicational universals are given in (3):

3) IMPLICATIONAL UNIVERSALS

(a) The presence of voiced stops implies the presence of voiceless stops but not vice versa (Jakobson, 1968: 70).

(b) The presence of voiced fricative implies the presence of voiceless fricatives, but not vice versa (Jakobson, 1968: 70).

(c) The presence of fricative implies the presence of stops, but not vice versa (Jakobson, 1968:51).

(d) The presence of aspirated stops implies the presence of unaspirated stops, but not vice versa (Jakobson, 1968: 51).

(e) The presence of palatalized consonants implies the presence of non-palatalized consonants (Fellbaum, see discussion).

(f) The maintenance of a voice contrast word-finally implies the maintenance of voice contrast word-medially, and the maintenance of a voice contrast word-medially implies the maintenance of a voice contrast word-initially, but not vice versa (Dinnsen & Eckman, 1975; in Eckman, p. 322).

(g) Aspiration rules that apply under certain conditions syllable-initially in word-medial syllable also apply under at least all the same conditions syllable-initially in word-initial syllables, but not necessarily vice versa
                                                        (Houlihan, 1977).

(h) The presence of a voiced stop word-finally implies the presence of a voiceless stop word-finally, but not vice versa (based on Dinnsen & Eckman, 1975).

Given these implicational relationship, the relative markedness of specific elements can be established as in (4):

(4) MARKEDNESS RELATIONS

(a) Voiced stops are marked with respect to voiceless stops.

(b) Voiced fricatives are marked with respect to voiceless fricatives.

(c) Fricatives are marked with respect to stops.

(d) Aspirated stops are marked with respect to unaspirated
     stops.

(e) Palatalized consonants are marked with respect to non-
     palatalized consonants.

(f) A word-final voice contrast is marked with respect to a
     word-medial contras, and a word-medial voice contrast is
     marked with respect to word-initial voice contrast.

(g) Aspirated stops syllable-initially in word-medial position
     (under certain conditions) are marked with respect to
     aspirated stops that appear syllable-initially in word-
     initial positions (under at least all the same conditions.)

(h) Word-final voiced stops are marked with respect to word-
     final voiceless stops.

Notice that the structures of each types of statement in 3 and 4
are not the same. For example, the first type, exemplified in a
through e are context-free and compare only corresponding features
or elements across languages, independent of the context in which
they occur (referred to here as Type I). The second type, f, g, and
h are referred to as context-sensitive implicational universal and
can be divided into two classes: 1) those which compare the same
phenomena in different environments (f and g) (Type II); and 2)
those which compare different phenomena in the same environment (h)
(Type III).

The three types of implicational relations with their respective
marking relations for voice contrast are given in 3 and 4, statements
a, f, and h. All three types of implicational statements are important for correct application of the MDH. Determining where a language maintains a voice contrast, using Type II, is necessary for explaining certain facts about the markedness relationship of a given
segment. For example, not only are voiced obstruents marked but
they are more marked distributionally in word final position than in
word initial position (See 4d). Hence the MDH would predict that
German speakers would have difficulty acquiring voiced obstruents in
word final position in English, whereas English speakers would not
have difficulty suppressing a more marked segment in the most marked
position for that segment. Eckman also provides another example of
a bi-directional learning situation, between English and French. In
French, the phoneme /ž/ occurs word-initially, -medially, and finally, whereas in English, it occurs only word-medially and finally.
The French learner of English must learn to suppress this contrast,
whereas an English speaker learning French must acquire this new
marked phoneme in word-initial position. Citing Gradman (1971),
Eckman pointed out that this learning situation is not difficult for

either the French or the English speakers. The MDH correctly predicts that this should be easy, since this is the least marked position for such a contrast, and hence should not be difficult (Eckman,
pp. 322-323).

To summarize, for making predictions using the MDH, first determine which segment is marked in a given differing pair of phones,
using Type I. Then determine the markedness of the specific environment in which they are found, using Type II. If the marked segment
is in the most marked position, it will be difficult to acquire. If
a marked segment is in the least marked position, it will not be
difficult to acquire. If a segment is unmarked, it should not be
difficult to acquire (produce) even if it is in the most marked
position on the hierarchy, as demonstrated by English speakers
learning German. This then establishes the direction of difficulty
between two segments in the same "area" of the grammar between two
languages (Eckman, p.318-323.) A chart which summarizes these three
types of implicational universals and the predictions the MDH makes
with respect to phonology is given in 5 (see next page).

English to Portuguese

Having completed a review of the basic notions used by the MDH,
let us turn to our three studies. The first is an experimental study
by Farber (1982) with English as the NL and Portuguese as the TL; its
results confirmed a variety of observational studies (Egon, 1981;
Stockwell & Bowen, 1965, & Hans Wolff, 1950.) The segments tested
were the voiceless stops /p, t, and k/. Both English and Portuguese
have all three stops in their phonemic inventory. In English these
stops alternate, occurring as aspirated and non-aspirated allophones.
Aspirated stops occur word-initially, independent of stress, and
word-medially in stressed syllables. The rule in 6 is a formal
statement of the process of English aspiration (excluding free
variation, which we will not discuss in this paper).

6)   Aspiration of voiceless stops in English:

$$
\begin{bmatrix} -son \\ -cont \\ -d.r. \\ -vce \end{bmatrix} \longrightarrow [+asp] \quad / \quad \left\{ \begin{matrix} \# \\ \$ \end{matrix} \underline{\hspace{1.5cm}} \underset{[+sts]}{V} \right\}
$$

Portuguese, on the other hand, shows no allophonic variation of aspiration in its series of voiceless stops anywhere in the grammar; in
other words, voiceless stops are unaspirated in all positions in Portuguese. Data from English and Portuguese demonstrate the difference
in the way these segments function between the two languages (Farber,
p. 15):

(5) SUMMARY OF MARKEDNESS RELATIONS OF THREE TYPES OF IMPLICATIONAL
    UNIVERSALS

PHONOLOGY

Voice contrast in Word-Final Position vs.
No Voice Contrast in Word-Final Position

Difference: German (Russian, etc.) has no word-final voice contrast
           and allows only voiceless obstruents word finally;
           English (French, etc.) has word-final voice contrast and
           thus allows both voiced and voiceless obstruents
           word-finally.

Learning   German speakers learning English (GS in chart);
Situation: English speakers learning German (ES in chart).

Type			Language Example	Implication	Markedness Relation	MDH Prediction of Difficulty
A. Type I						
$\begin{bmatrix}+obst\\+vce\end{bmatrix}$	$\begin{bmatrix}+obst\\-vce\end{bmatrix}$		English, German	$\begin{bmatrix}+obst\\+vce\end{bmatrix} \blacktriangleright \begin{bmatrix}+obst\\-vce\end{bmatrix}$	vcd obst is more marked than voiceless obst.	GS: YES ES: NO
* $\begin{bmatrix}+obst\\+vce\end{bmatrix}$	$\begin{bmatrix}+obst\\-vce\end{bmatrix}$		Hawaiian			
$\begin{bmatrix}+obst\\+vce\end{bmatrix}$	$\begin{bmatrix}+obst\\-vce\end{bmatrix}$		None			
* $\begin{bmatrix}+obst\\+vce\end{bmatrix}$	$\begin{bmatrix}+obst\\-vce\end{bmatrix}$		None			
B. Type II						
$\begin{bmatrix}+obst\\+vce\end{bmatrix}/\_\#$;	$\begin{bmatrix}+obst\\-vce\end{bmatrix}/\#\_$		English	$\begin{bmatrix}+obst\\+vce\end{bmatrix}/\_\# \blacktriangleright \begin{bmatrix}+obst\\+vce\end{bmatrix}/\#\_$	vc. contrast word-finally is more marked than vcd. contrast word-initially	GS: YES ES: NO
* $\begin{bmatrix}+obst\\+vce\end{bmatrix}/\_\#$;	$\begin{bmatrix}+obst\\-vce\end{bmatrix}/\#\_$		German			
$\begin{bmatrix}+obst\\+vce\end{bmatrix}/\_\#$;	* $\begin{bmatrix}+obst\\-vce\end{bmatrix}/\#\_$		None			
* $\begin{bmatrix}+obst\\+vce\end{bmatrix}/\_\#$;	* $\begin{bmatrix}+obst\\-vce\end{bmatrix}/\#\_$		Rapanui			
C. Type III						
$\begin{bmatrix}+obst\\+vce\end{bmatrix}/\_\#$;	$\begin{bmatrix}+obst\\-vce\end{bmatrix}/\_\#$		English	$\begin{bmatrix}+obst\\+vce\end{bmatrix}/\_\# \blacktriangleright \begin{bmatrix}+obst\\-vce\end{bmatrix}/\_\#$	vc. obst. word-finally is more marked than voiceless obst. word-finally	GS: YES ES: NO
* $\begin{bmatrix}+obst\\+vce\end{bmatrix}/\_\#$;	$\begin{bmatrix}+obst\\-vce\end{bmatrix}/\_\#$		German			
$\begin{bmatrix}+obst\\+vce\end{bmatrix}/\_\#$;	* $\begin{bmatrix}+obst\\-vce\end{bmatrix}/\_\#$		** Dakota			
* $\begin{bmatrix}+obst\\+vce\end{bmatrix}/\_\#$;	* $\begin{bmatrix}+obst\\-vce\end{bmatrix}/\_\#$		Rapanui			

NOTE: -Capital letters in MDH prediction of difficulty indicated those predictions
      which are supported by evidence, i.e., correct predictions by the MDH.
      -The phonology example is a compilation of Gundel and Eckman's discussion of
      this phenomenon, with examples provided in part by the author.

**The situation in Dakota is somewhat more complex than this chart permits.
For a discussion of all the conditions as well as of actual occurrences of
this phenomenon see Steyaert, 1978.

7) Word Initial Voiceless Stops:

Portuguese Data			English Data	
peca	[pɛsa]	'play'	positive	[pʰazətɪv]
taca	[tasa]	'champagne glass'	tonic	[tʰanIk]
capa	[kapa]	'cape'	copper	[kʰapɚ]
panela	[panɛla]	'pot'	capacity	[kʰpʰæsIti]
Caribe	[karibɾ]	'Caribbean'		

Farber's study researched aspiration phenomena in both learning directions, Portuguese to English and English to Portuguese. She tested four native speakers, two from each language. The data collected of Portuguese speakers learning English found that Portuguese speakers had difficulty acquiring the aspirated stop in English. Since our paper is concerned with the acquisition of non-alternating unmarked segments, we will confine ourselves to Farber's discussion of English speakers learning Portuguese. (The native English speakers were two female speakers of the Minnesota dialect of English. Each was from either the middle or upper class and had at least one year of college education. Both were advanced students of Portuguese, having studied Portuguese at least eight years and were judged to be approximately equal in their spoken fluency.)

The voiceless stops, /p, t, and k/ were tested in both syllable-initial and syllable-final positions, including word-initially with stressed and unstressed vowels. The test items were divided into three categories: 1) nonsense words; 2) Portuguese words; and 3) English words. Nonsense words were placed in a carrier phrase in the subjects' native language; the English tokens were placed in the carrier phrase: "Say the word _____ again".

In the study Farber found that the values for VOT for voiceless stops in word initial position were very similar in both the Portuguese words and the English words for the native English speakers. She also found that VOT was significantly greater for /p,t,k/ produced in the Portuguese words by the native English speakers than it was for the same stops produced by native speakers of Portuguese also tested. In other words English speakers had a difficult time learning to produce voiceless unaspirated stops word-initially in Portuguese.

Icelandic to English or German

The data from Icelandic speakers is taken from Thraínsson, 1978 and involves two target languages, English and German. The segments to be examined are the preaspirated voiceless stops [hp, ht, and hk]. One area of the Icelandic grammar in which preaspirated stops occur is in syllables with short vowels when these consonants are geminated (See Note 1). The following data from Thraínsson illustrate (Y = u):

8) Adjectives

Fem. Sg.	Neut. Sg.	Gloss
[rík]	[rík+t]	'rich'
[djúp]	[djúp+t]	'deep'
[fei:tʰ]	[feiht]	'fat'
[ljou:tʰ]	[louht]	'ugly'
[sai:tʰ]	[saiht]	'sweet'

According to one of Thraínsson's analyses, the voiceless stops
/p, t, and k/ become [h] before an identical stop or in geminate
clusters in syllables with a stressed vowel. Icelandic contains no
syllables with a stressed long vowel followed by a long consonant;
therefore these voiceless stops can not occur as geminates in syllables with long vowels. Instead, they are found in words with a
stressed short vowel, where a long consonant (two or more) must
follow. The pre-aspirated segments thus occur in an environment
with a stressed short vowel, i.e., in the environment where long
consonants occur in this language. When this long consonant is /p,
t, or k/, the stop undergoes the preaspiration rule stated below
(Thraínsson, p. 33)

9) Preaspiration in Icelandic

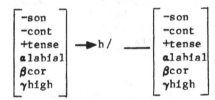

(The feature [+ tense] is used to distinguish p, t, k from b, d,
g.)

When Icelanders are faced with a stressed VC sequence in foreign
languages, they reinterpret them as either a long vowel followed by
a short consonant or short vowel followed by a long consonant, V:C
or VC:, respectively. In German which has relatively clear differences between short and long vowels, it is quite simple: words with
short vowels are analyzed as containing a vowel followed by a phonologically long or geminate consonant as shown below. When the consonant is either [pp, tt, or kk], the preaspiration rule applies as in
(10a). When the vowel is long, on the other hand, the succeeding
consonant is analyzed as short, shown in (10b), fitting in correctly
with both the German and Icelandic system, so preaspiration does not
occur in the interlanguage. (IL). Words which illustrate the process of preaspiration in German by Icelandic speakers are given
below (where //ɹ..// = reinterpretation of IL; * = ungrammatical
sequence) (Thraínsson, p. 13):

10)   German   Icelander   Gloss

     a)[lIpə]   /lIppə/   *[lihpɛ]   'Lippe'   'lip'
       [mIt[/mItt/   *[mIht]   'mit'   'with'
       [mItə]   /mIttə/   *[mIhtɛ']   'mitte'   'middle'
       [akər]   /akkər/   *[ahkər]   'acker'   'field'
     b)[mu:t]   /mu:t /   [mu:t]   'Mut'   'courage'
       [mi:tə]   /mi:tə /   [mi:tə]   'Miete'   'rent'

In English, on the other hand, which does not have long and
short vowels phonemically, the "lax" vowels such as [I, ,] in
stressed positions are generally interpreted as short by Icelandic
students learning the language. In these cases the preaspiration
rule applies as in (11a). In words with tense or dipthongal vowels,
however, the consosnants are analyzed as short and correctly pronounced without preaspiration given in (11b) (Thraínsson, p. 13):

11)   English   Icelander   Gloss

   a) [rIp]   //ripp//   *[rIhp]   'rip'
       [mɛt]   //mɛtt//   *[mɛht]   'met'
       [lItɛr]   //lIttɛr//   *lIhtɛr]   'litter'
       [khɔ pər]   /khɔppɛr//   *[khɔ hpɛr]   'copper'
       [sɔkɛr]   //sɔkkɛr//   *[sɔhkɛr]   'soccer'
       [lɔtɛri]   //lɔttɛri// *[lɔhtɛri]   'lottery'
   b) [dIph]   //dI:ph //   [dI:ph]   'deep'
       [feth]   //fe:th //   [fe:th]   'fate'
       [kokh]   //ko:k h//   [ko:kh]   'coke'

The above data of native speakers of Icelandic learning German
and English, respectively, shows difficulty in acquiring the unaspirated voiceless stops of the two target languages. These surface
in a "similar" (re-analyzed) environment in the interlanguage where
they occur in their native language, after stressed short vowels.
These preaspirated stops fall under a general definition of aspiration in Ladefoged (1971), where voicing lag is defined as "a period
of voicelessness." (In preaspirated stops this lag precedes stop
closure; in post-aspirated stops it follows stop closure. Using
this same definition, Houlihan (1977) found that preaspirated stops
follow the same generalizations and fall into the same classificatory types as post-aspirated stops typologically, an example of
which was given above in 3g. There thus appears to be evidence for
the operational validity of Ladefoged's definition of aspiration for
typological and universal language studies (See Note 3). If we
apply this same definition to the implicational universal given in
3d, as well, Jakobson's relation of aspirated to unaspirated stops
would include both pre- and post-aspirated stops. This gives us the
markedness relation found in (4d) with pre-aspirated stops, a subset
of aspirated stops, being marked with respect to unaspirated (or

non-aspriated stops.). Thus, we see that native speakers of Icelandic have difficulty acquiring the less marked unaspirated stops of German and English which correspond to a marked segment in the same (or similar) environment in their native language.

Polish to English

The data presented in this section, Polish students learning English, is taken from Rubach (Rule Typology and Phonological Interference (forthcoming). Although several palatalization rules exist in Polish, we will consider only the rule of Surface Palatalization (See Note 4). In this process, the feature of palatalization is superimposed on a consonant without changing the place of articulation. (See Note 5). It affects all consonants, morpheme internally and at word boundaries, whenever they are followed by the segments /i/ or /j/. This phenomenon is exemplified by the following morphemes (Rubach, p. 12):

POLISH	GLOSS
[p'iasek]	'sand'
[row'i]	'ditch and'
[t'ik]	'twitch'
[l'ista]	'list'
[z'jazd]	'convention'
[k'ino]	'cinema'
[gloš' jego]	'his voice'

Rule 9 is Rubach's, formal statement of this process of palatalization (p. 11):

12) Surface Palatalization

$$[\text{+cons}] \longrightarrow \begin{bmatrix} \text{+hi} \\ \text{-bk} \end{bmatrix} \quad / \quad \underline{\hspace{2cm}} \quad ([\text{-seg}]) \quad \begin{bmatrix} \text{-cons} \\ \text{+hi} \\ \text{-bk} \end{bmatrix}$$

The following data from Rubach show evidence of this process occuring in the speech of Poles learning English (p. 12):

13)

English	Poles	Gloss
[pik]	*[p'ik]	'peak'
[šild]	*[s'ild]	'shield'
[tim]	*[t'im]	'team'
[lik]	*[l'ik]	'leak'

Thus, Polish speakers learning English, incorrectly produce palatalized consonants in those environments in English where consonants are palatalized in Polish. Since some languages, e.g. Hawaiian, have

only non-palatalized consonants with no palatalized variants, while
other languages, such as Polish and Russian, have both palatalized
and non-palatalized consonants, it seems reasonable to conclude that
the presence of palatalized consonants implies the presence of non-
palatalized consonants, but not vice versa (3e). Therefore, palatalized consonants can be said to be marked with respect to non-palatalized consonants (4e). According to the MDH, the less marked, or
non-palatalized consonants, should be easy to learn for Polish students of English. However, invariably they pronounce palatalized
consonants, or the marked member of the pair, in those environments
in English where the nonpalatalized or less marked segment is found
in Polish.

As stated in the introduction, the MDH has met with few, if any,
problems in predicting relative ease or difficulty of producing the
correct unmarked phoneme in a TL, where a marked non-alternating
phoneme contrasts with this phonemic category in the second language
learner's NL (e.g., English speakers learning German). As demonstrated in the data presented in this paper, however, it is not as
successful in accounting for the facts about marked phones (e.g.,
[p^h]), which act as allophonic variants of the phonemic category
in a NL system, but as non-alternating unmarked phonemes (e.g., /p/)
in a TL, although the segments in both interactions may differ only
in one feature, i.e. [+vce] vs. [−vce] or [+asp] vs. [−asp], etc.
Furthermore, implicational universals which characterize only binary
distinctions of similar segments in the same context can not offer a
solution to this difference in interaction either.

But phonemes and allophones differ in more ways than simply the
feature specifications of the segment as far as the phonological system of a language is concerned. Distinctive features are those features which distinguish one sound from another, and they characterize
the phones which interact as phonemes in positions of contrast. Non-
distinctive features, on the other hand, do not distinguish one
sound from another; they are redundant features, predictable from
the context, in complementary distribution with the other allophonic variant. It has been assumed by linguists (e.g., Hans Wolff,
1950; Weinreich, 1953; Lado, 1957; Koutsoudas & Koutsoudas, 1962;
Stockwell & Bowen, 1965, to mention only a few), "that it is this
very existence of a system of distinctive and non-distinctive
features which causes interference when the speaker of one language
attempts to learn another language in which the phonological system
is composed of partially similar and completely different distinctive and non-distinctive features" (Brière, p. 15). It is this net-
work of difference between sounds which must therefore be captured
in order to adequately account for "those element that are similar
to [one's] native language...and those elements that are different"
in which Lado had stated lay "the key to ease or difficulty in foreign language learning." A theory of second language acquisition
which appeals to a theory of markedness for its predictive state-

ments must be able to account for this interaction. Table 1 summarizes some of the key facts in each of the language interactions
presented in this paper, as well as those discussed by Eckman.

This chart uses four parameters of comparison. The first column,
labelled "Phones", shows the particular segments from each native
language compared to the same segment in the same environment in the
target languages; we have only listed the segments which change in
the respective target languages discussed. For example, English
speakers must learn not to produce (suppress) voiced stops word-
finally in German. For German language learners, a new, voiced
phone must be acquired in the environment being compared between the
two languages. In the last five language groups, each of the
segments in the native language already has a counterpart in the
target language. The learner must learn to alter the segment from
his native language, rather than add or delete a segment which
occurs in the respective enviroment being compared.

The column labelled "Environment Conditions" categorizes the
segments according to whether or not the particular phone has a restricted, or contextually conditioned distribution, or whether it
occurs with no restrictions in the native or target language. For
example, in English, voiced and voiceless phones occur in all positions in a word, including word-finally. In German, the voiceless
phone does not occur word-finally, so it cannot be contextually
altered. In those language groups which have a contextual constraint on the segments discussed, the conditions are given. In
Greek the segments occur freely and it is the target language which
has the contextual restrictions. In the last four language groups,
the native languages all have the segments occurring freely.

The third column, "Contrast," is provided to show the difference
between the first two language-learning situations and all the rest.
The segments do not undergo alteration, but rather addition and deletion of segments. In the first case, English to German, a contrast
is neutralized; in the second, a contrast must be acquired in a new
environment. In the other five situations, a contrast exists in both
the native and target languages. Segments are not added or deleted,
but rather features are added or deleted, preserving the contrast in
both languages involved. It is this alteration of the original cha-
racter of the phone which seems closer to what Koutsoudas and Koutsoudas were referring to as similar. Eckman and Gundel seem to be
referring to the acquisition or deletion (labelled suppression by
Eckman) of segments.

The last column shows the markedness of the individual segments,
determined by context-free implicational statements, exemplified in
3a and 3b. The markedness of the segments for German and English
can also be determined by using the context-sensitive implicational
universals based on the voice contrast typology from Dinnsen and
Eckman (3e). For each of the other cases there is no markedness

Table 1. Summary of Language Data.

		*PHONES		ENVIRONMENT CONDITIONS		CONTRAST		MARKEDNESS OF PHONES	
NL	TL	NL	TL	NL	TL	NL	TL	NL	TL
A. English	German	b	p	free	free	yes	no	M	U
German	English	-	b	N/A	free	no	yes	-	M
B. Greek	English	p	p^h	free	#___	yes	yes	U	M
English	Portuguese	p^h	p	#___	free	yes	yes	M	U
Polish	English	C'	C	___$\left\{\begin{smallmatrix}i\\y\end{smallmatrix}\right\}$	free	yes	yes	M	U
Icelandic	German	hp	p	V́CC	free	yes	yes	M	U
Icelandic	English	hp	p	V́CC	free	yes	yes	M	U

*p=voiceless stops p, t, and k; b=voiced stops b, d, and g.
C=consonant; C'=palatalized consonant.

relationship in the respective contexts for the segments. For example, aspirated stops word-initially do not imply unaspirated stops
word-initially, or vice versa, therefore, there is no markedness
relation between aspirated and unaspirated stops in this environment.

The data from these three language learning situations do not
fit the predictions of the MDH, as currently worded. Either the
MDH, as currently formulated, makes wrong predictions or it does not
apply to these language learning situations. If the facts about
each of these acquisition processes all possess common characteristics, then it may be possible to formulate a general statement to
account for the apparent discrepancy between these learning situations and the predictions of the MDH.

The four native language learning situations in Part B differ
from those of Part A in three main ways: (1) they all have a marked
segment in the same environment where an unmarked segment appears in
the TL; (2) the distribution of the NL segment is predictable from
context in one language whereas in the TL no restrictions on the
distribution of this segment exist; (3) no context-sensitive implicational universal (Type III) exists which directly relates the segments in the given environment between the two languages involved;
e.g. word-initial aspirated [ph] cannot be more marked than word-
initial unaspirated [p^h] since English has the former and Portuguese has the latter.

Since a general pattern of interaction between the respective NL and TL language learning situations exists, the following revision is suggested to incorporate all the language situations discussed in this paper:

PROPOSED REVISION TO THE MARKEDNESS DIFFERENTIAL HYPOTHESIS

The areas of difficulty that a language learner will have can be predicted on the basis of a systematic comparison of the grammars of the native language, the target language and the markedness relations stated in universal grammar, such that,

(a) Those phenomena in the target language which differ from and are typologically more marked than corresponding phenomena in the same context in the native language will be difficult.

(b) Those phenomenon which differ from and are typologically less marked than corresponding phenomena in the same context in the native language will not be difficult.

(c) Those phenomena in the target language which differ from and are not more marked than the corresponding phenomena in the native language will be difficult to learn, if they are contextually predictable in the native language but not contextually predictable in the target language.

Although this paper has only concentrated on phonological data, similar interactions of marked and unmarked segments, distributional facts, and morphemes versus allomorphic variants are patterns of the interactions between the syntax of a NL and TL. This revised clause still needs to be applied to the acquisition of syntax to see if the same generalizations hold true. If the proposed hypothesis does explain the acquisition of syntax as well as phonology it would maintain the position of Eckman's use of markedness and universal statements from grammar to explain the difficulties in second language acquisition.

Notes

Acknowledgments: The author would like to thank the following people for their comments in the preparation of this paper: Stig Eliasson, Jeanette Gundel, Larry Hutchinson, and Fred Eckman. I would especially like to thank Monika Forner for her willing ear and Gerald Sanders for his patient and untiring discussion and relentless questioning. Of course any errors found in this article are solely my own.

1. Preaspiration also occurs before /p,t,k/ + /l, m, n/. We

have selected Rule 9 due to the availability of the second language
data in his portion of this article, lacking in his other discussions. Data for the other environments give evidence of alternation
in the segments /p, k/ which is not evidenced in the data in 12.
(Thraínsson, p. 17.)

2. Thraínsson points out that 1) not all speakers pronounce the
feminine form with an aspirated /t/; 2) "It is sometimes claimed
that all stops are post-aspirated in final position in Icelandic.
The reason for this claim is that it is very difficult to perceive
any difference in postaspiration between words like 'ljott' and
'ljott' for instance, in utterance-final position. But such differences are clear if a vowel-initial word follows, cf. 'ljota litinn
'of ugly color' with postaspiration vs. 'ljott a litninn' with preaspiration and no postaspiration. Hence I shall mark postaspiration
on words like ljot, saet, feit, etc., but not on 'ljott', 'saett',
'feitt', or the like." We mention these facts here in order to
clarify that the process of preaspiration in Icelandic should not be
analyzed as interrelated to an existing post-aspirated segment. The
two processes are clearly distinct in the grammar.

3. Thraínsson's analysis of this process in Icelandic states
that "preaspiration is not simply the inverse of postaspiration, as
it name and some phonetic description might lead us to believe", but
rather he maintains a rejection of the view sometimes advanced that
preaspiration is to be considered a "component or a phonetic feature
of the succeeding stop (cf. 3.g. Haugen 1958:72)". He thus tran-
scribes postaspirates above the line [h], while giving full segment
status to preaspiration (p. 5). His rejection of this view is due
especially to his preference of a phonetic and phonological explanation of a "spreading of the glottis with no inherent supralaryngeal
configuration (p. 5).

4. Rubach's discussion includes several rules in Polish which
gives rise to palatal consonants: (1) 2nd Velar Palatalization, a
morphologically-conditioned rule which palatalizes velars before
front vowels in a certain class of suffixes (formalized in Rubach,
forthcoming, Palatalization, cyclic phonogloy, and borrowings into
Polish and English); (2) Anterior Palatalization which changes
dental consonants into prepalatals before front vowels; (3) Nasal
Palatal Assimilation in which a nasal consonant assimilates to the
point of articulation of a following prepalatal nasal; (4) Strident
Assimilation which assimilates dental fricatives to the point of
articulation of certain nonanterior noncontinuants. Of these, only
rule 3, Nasal Palatal Assimilation, also gives rise to transfer in
the interlanguage of Poles learning English. This rule of nasal
palatalization applies exclusively to nasal consonants in all positions in the word and at word boundaries, as reported by Rubach. I
have chosen not to include it in this discussion, because influence
from English orthography appears to be a strong complicating factor.

In Polish the rule is triggered whenever a nasal consonant is followed by a high front vowel, and this is indicated in the orthographic representation. Rubach gives only English words which fit exactly the orthographic representation in Polish, e.g. "minimum,"
"nirvana", and "training". Further research needs to be done to
verify the phonological difficulty of English rather than these
orthographic complications.

5. In the case of the velars /k, g, x/, all [+hi] in the input
to the rule, Surface Palatalization does affect the place of articulation in that it fronts these consonants from velars to postpala-
tal (prevelar) position." (Rubach, forthcoming, p. 11 of unpublished manuscript).

References

Bhat, D.M.S. "A General Study of Palatalization." Universals of
    Human Language, Vol. II, Ed. Mouton, 1963, pp. 47-92.
Brière, Eugene John. A Psycholinguistic Study of Phonological Inter-
    ference. The Hague: Mouton &Co. N.V. 1968.
Dinnsen, Daniel and Fred Eckman. "A Functional Explanation of some
    Phonological Typologies." In R. Grossman et al, eds.
    Functionalism, CHicago Linguistic Society, 1965, pp. 126-134.
Eckman, Fred R. 1977. Markedness and the Contrastive Analysis
    Hypothesis. Language Learning. 27. 315-330.
Egon, Veronica. 1981. Eckman's Markedness Differential Hypothesis:
    It's Viability in English and Spanish. Unpublished University
    of Minnesota manuscript.
Eliasson, Stig. (ed.) Forthcoming. Theoretical Issues in Contras-
    tive Phonology. (Studies in Descriptive Linguistics 13.)
    Heidelberg: Julius Groos Verlag.
Farber, Beth. 1982. The Voiceless Stop Consonants of Portuguese
    and English: A Contrastive Analysis. Unpublished U of Minnesota
    manuscript.
Gradman, H. 1970. The Contrastive Analysis Hypothesis: What it is
    and what it isn't. Ph.D. dissertation, Indiana University.
Gundel, Jeanette Kohn. 1982. "Another Look at the Markedness Dif-
    ferential Hypothesis." Paper presented at the 11th Annual U of
    Wisconsin-Milwaukee Linguistics Symposium on Universals of
    Second Language Acquisition, March 19 & 20, 1982.
Houlihan, Kathleen. On Aspiration and Deaspiration Processes.
    Current Themes in Linguistics: Bilingualism, Experimental
    Linguistics, and Language Typologies, ed. F.R. Eckman, 215-
    239. Washington, D.C.: Hemisphere Publishing Corporation.
    1977.
Jakobson, R. 1940. Kindersprache, Aphasie, und Allgemeine Lautge-
    setze. (Tr. as Child Language, Aphasia and Phonological Uni-
    versals. The Hague: Mouton, 1968.)

Koutsoudas, Andreas and Olympia Koutsoudas. "A Contrastive Analysis
    of the Segmental Phonemes of Greek and English." Language Learn-
    ing, 12, (1962) 211-230.
Lado, R. 1957. Linguistics Across Cultures. Ann Arbor, Michigan:
    The University of Michigan Press.
Lisker, Leigh and Arthur S. Abramson. "A Cross-Language Study of
    Voicing in Initial Stops: Acoustical Measurements." Reprinted
    in Word 20, No. 3, (1964) pp. 385 - 422.
Rubach, Jerzy. Forthcoming. Rule Typology and Phonological Inter-
    ference. To appear in Eliasson, forthcoming Theoretical Issues
    in Contrastive Phonology. (Studies in Descriptive Linguistics
    13.) Heidelberg: Julius Groos Verlag.
Sanders, G. 1977. A Functional explanation of elliptical coordi-
    nations. In F. Eckman (ed.), Current Themes in Linguistics.
    Washington, D.C.: Hemisphere Publishing Company.
Smalley, William A. 1973. Manual of Articulatory Phonetics.
    Revised Edition. William Carey Library, South Pasadena.
Stevens, Kenneth N. 1975. Modes of Conversion of Airflow to Sound
    and their Utilization in Speech. Paper presented at the Eighth
    International Phonetic Sciences, Leeds, England, August 17-23,
    1975.
Stockwell, Robert P. & J. Donald Bowen 1965. The Sounds of English
    and Spanish. Chicago: University of Chicago Press, pp. 7-18.
    Reprinted in Robinett, Betty Wallace & Jacquelin Schachter, eds.
    Second Language Learning: Contrastive Analysis, Error Analysis,
    and Related Aspects, "Sound Systems in Conflict: A Hierarchy of
    Difficulty", (1983) Ann Arbor, Michigan: The University of
    Michigan Press, pp. 20-31.
Viana, Maria de Ceu. "O Indice Duracao e a Analise Acoustics das
    Oclusivas Orais em Portugues." Boletim de Filologia, Vol. XXV.
    (1979) Lisbon.
Wolff, Hans, "Partial Comparisons of the Sound System of English and
    Puerto Rican Spanish", Language Learning, III, 1 and 2 (1958),
    pp. 38-418.
Weinreich, Uriel. 1953. Languages in Contact. New York: Publica-
    tions of the Linguistic Circle of New York.
Zlatin, Marsha A. 1974. Voicing contrast: perceptual and productive
    voice onset time characteristics of adults. Journal of the Acou-
    stical Society of America 56. pp. 981-94.

MARKEDNESS AND PARAMETER SETTING: SOME IMPLICATIONS

FOR A THEORY OF ADULT SECOND LANGUAGE ACQUISITION

Lydia White

McGill University

1. Introduction

The concept of markedness has traditionally been invoked to try
to provide a description and explanation of aspects of language that
are are felt to be 'unnatural', infrequent, complex or lacking gener-
ality. In the field of first language acquisition, it is usually
assumed that marked forms or structures, however defined, are harder
to learn and are acquired after unmarked (e.g. Chomsky 1969; Clark
1970; Kiparsky 1974; Phinney 1981). Markedness has also been used
to predict or explain certain difficulties that second language
learners experience (Eckman 1977), as well as the order in which
they acquire certain constructions, with unmarked preceding marked
(hyltenstam 1982; Rutherford 1982, Mazurkewich 1984a).

In recent linguistic theory, attempts to characterize markedness
in the syntactic domain have centered round the concept of 'core
grammar' (Chomsky and Lasnik 1977; Chomsky 1980, 1981a, 1981b, 1982):

"UG consists of a highly structured and restrictive system of
principles with certain open parameters, to be fixed by experience. As these parameters are fixed, a grammar is determined,
what we may call a 'core grammar'" (Chomsky 1981b, 38).

Markedness operates at two levels, in that some parameters of UG may
be fixed before others, or certain settings of certain parameters
may be assumed before others, and, in addition to core grammar,
there is a marked periphery of more language-specific rules:

"(In) our idealized theory of language acquisition, we assume
that the child approaches the task equipped with UG and an

associated theory of markedness that serves two functions: it
imposes a preference structure on the parameters of UG, and it
permits the extension of core grammar to a marked periphery.
Experience is necessary to fix the parameters of core grammar.
In the absence of evidence to the contrary, unmarked options are
selected" (Chomsky 1981a, 8).

The child's initial hypothesis constitutes the unmarked case, the
assumption that he or she will make in the absence of counter-evidence, whereas a marked option requires specific evidence and will
not be considered without such evidence.[1]

When one considers the second language learner, the definition
of marked forms as those requiring specific evidence raises interesting questions, since second language learners are faced with evidence from two different languages. They must distinguish which
evidence is relevant for which language. In some cases, the first
language (L1) and the second language (L2) will have parameters of
UG set in different ways.

In this paper, I shall explore the possibility that adults learning a second language have problems resetting the parameters of UG,
particularly when marked parameters or marked peripheral rules have
already been established for L1 but are inappropriate for L2.   In
these circumstances, it may prove difficult or impossible to 'unset'
the marked parameter. That is, I suggest that interference from the
mother tongue is particularly likely to occur in two circumstances:

i.   where the learner carries over a marked construction from
      L1 to L2

ii.   where the learner incorrectly assumes that some setting of
      a parameter of UG is relevant for L2, so that a range of
      apparently unrelated errors can be accounted for, as would
      be the case with incorrect choice of bounding nodes for
      subjacency, for example.

2.   Recent claims on markedness and second language acquisition

A number of researchers in recent years have been interested in
markedness as a source of explanation in second language acquisition.
Rutherford (1982) reviews work in this area and suggests that the
main trend has been to apply the notion of markedness to cases of
L1-L2 transfer, for example to characterize what kinds of phenomena
are or are not subject to transfer. He suggests that there has been
little systematic attempt to apply markedness claims to developmental studies of second language acquisition, that is, to show that
some aspects of L2 acquisition are characterized by the emergence of
unmarked structures before marked, as is usually supposed for L1
acquisition.[2]

However, looking closely at many recent studies, one finds that
they contain the seeds of a developmental approach, since the general
assumption is made that the L2 learner will first adopt the unmarked
case in L2, _irrespective_ of his L1 experience. That is, a number of
researchers imply or explicitly claim that unmarked forms are easier
and will be tried first by the second language learner (Eckman 1977;
Ioup and Kruse 1977; Gass 1979; Mazurkewich 1984a). Hyltenstam
(1982) summarizes these claims as follows:

Table 1

Native	Language (L1)	Target	language (L2)	Initial	interlanguage
i.	unmarked		unmarked		unmarked
ii.	unmarked		marked		unmarked
iii.	marked		unmarked		unmarked
iv.	marked		marked		unmarked

Since the form in the initial interlanguage is unmarked in all
cases, language transfer cannot be said to be a factor at all.
Rather, the assumption is that all language learning, whether of L1
or L2, follows the order 'unmarked' before 'marked', regardless of
the data available to the learner. For L2 acquisition, this means
that the mother tongue will have no effect, and that certain aspects
of the target language data, namely the marked aspects, are
initially ignored.

I shall argue, on the contrary, that when markedness is considered in terms of parameters of UG or structures requiring specific
positive evidence, there will be times when L1 does have an effect.
In particular, the order of acquisition will not invariably be
'unmarked' before 'marked'. I will also suggest when it is possible
for the learner ultimately to achieve success in L2 and when success
is unlikely.[3]

3. Some predictions as to the effects of markedness in L2 acquisition

The predictions to be discussed here are summarized in Table (2),
below. They are made on the assumption that the form adopted in the
initial interlanguage depends not only on markedness considerations
but also on an interaction between the kinds of evidence available
in each language. In other words, there may be transfer effects,
but only in certain circumstances:

As can be seen by comparing Tables (1) and (2), there are certain
cases where the claims made here coincide with previous views, as
well as situations where they differ considerably. In addition,
certain situations are discussed which have not explicitly been con-

Table 2

	native language (L1)	target language (L2)	early interlanguage	later interlanguage
i.	U	U	U	U
ii.	–	U	U	U
iii.	U	M	U	M
iv.	U and M	U and M	U	U and M
v.	X	Y	X	Y
vi.	–	M	M	M
vii.	M	U	M	M
viii.	M	M	M	M

sidered in the markedness framework but which nevertheless seem relevant. I shall comment below on each of the situations summarized in Table (2), discussing evidence and counter-evidence.

i. Where both languages show the unmarked setting of a parameter, this setting should show up in the L2 interlanguage at all stages, since neither L1 nor L2 furnish the necessary evidence for a marked form. Given the definition of markedness as the requirement for specific positive evidence, one would not expect marked forms to show up spontaneously.

ii. Where L1 has not had a particular parameter set at all, the unmarked setting should be triggered for L2 learners on the basis of exposure to an L2 which shows the unmarked setting. Since L1 has contributed nothing of relevance to the parameter in question, there will be no interference. Ritchie (1978a) shows that Japanese adults learning English as a second language constrain rightward movement rules, which are non-existent in Japanese, in terms of the Right Roof Constraint (Ross 1967).[4]. In other words, Japanese adults observe a universal constraint although they have never before been exposed to data relevant to that constraint, suggesting that parameters of UG can still be set for adults on exposure to L2 data.

iii. Where L1 has the unmarked setting and L2 the marked, one would expect adult learners eventually to acquire the marked setting, since the evidence for it is clearly there in the L2 data. Initially, however, they may try the unmarked case, not necessarily because this is dictated by markedness, but because this is the setting they are familiar with in L1. Thus, the claims of traditional contrastive analysis and of markedness are indistinguishable here.

Mazurkewich (1984b) argues that markedness considerations, rather than interference, lead L2 learners first to adopt the unmarked case in their interlanguage. She investigates the acquisition of dative questions by teenage native speakers of French and Inuktitut learning English as a second language. Dative questions in English are of two kinds, as illustrated in (3a) and (3b):

3. a. To whom did John give the book____?
   b. Who did John give the book to ____?

(3a) involves 'pied-piping'; that is, the whole prepositional phrase is moved to the front of the sentence. (3b), on the other hand, involves preposition-stranding, so that only the WH-pronoun is moved. Preposition-stranding is usually considered marked (van Riemsdijk 1978; Hornstein and Weinberg 1981) and pied piping unmarked.[5]. In English, however, stranding is common and pied-piping rare. Mazurkewich found that her French subjects produced more sentences like (3a), the unmarked version with pied-piping, and that they showed a developmental sequence towards the marked form, with the more advanced students using more instances of preposition-stranding than the intermediate students, who, in turn, used more than the beginners. Since French has forms equivalent to (3a) but not (3b), these results could be due to interference from the first language. Therefore, the results from the Inuit students, whose language does not contain prepositions, are crucial to the claim that markedness is the determining factor in the acquisition sequence. As will be discussed below (section 3.vi), the results from the Inuit students do not clearly support the claim of a developmental sequence from unmarked to marked in the acquisition of dative questions.[6]

iv. Situations where both languages have unmarked and marked versions of a rule may provide the best means to distinguish between the effects of interference and those of markedness. If markedness is a crucial factor, then the unmarked case should initially predominate in the L2 interlanguage, with the marked acquired later. This would show that the L2 learner does indeed pay selective attention to certain aspects of the data, namely the unmarked aspects, instead of assuming that everything possible in L1 is also possible in L2. Kellerman (1979) discusses cases where learners reject marked forms in L2 in favor of unmarked, even though their L1 contains equivalent marked forms.[7]

v. In certain cases, parameters of core grammar may be set differently for different languages without there being any implication that one setting is more marked than another. For example, core grammar will allow a number of unmarked word-orders consistent with X' theory, where specific evidence will be required to determine which order is the correct one for the language in question. In such cases, the different settings of parameters in L1 and L2 might cause initial problems for the language learner. Evidence seems to be conflicting on this point: for example, Schachter (1974) and Flynn (1981) argue that Japanese and Chinese adults have difficulties associated with the difference in the branching direction of these languages as compared to English. Ioup and Kruse (1977) and Gass (1979), on the other hand, suggest that this is not a factor. Work by Jansen, Lalleman and Muysken (1981) on the acquisition of Dutch word-order by Turkish and Moroccan adults suggests that the L1 word-order does have an effect initially.

The three situations to be discussed below differ more radically
from the claims usually made for markedness in L2 acquisition, in
that it is argued here that there are cases where the L2 learner
immediately adopts the marked form in his interlanguage. The nature
of the positive evidence will be a crucial factor in these cases.

   vi. Where L1 does not involve a particular parameter at all, or
does not have some rule, but L2 has the marked form, I suggest that
the L2 learner will not necessarily go through a stage of first acquiring the unmarked version. This is contrary to the claim of
Mazurkewich (1984b), discussed above. In the case of questions like
(3a) and (3b), she argues that the sequence shown by the Inuit reflects that of her French subjects who prefer the unmarked version
to the marked. However, as she herself notes, the Inuit subjects at
all levels (beginners, intermediate and advanced) consistently produce a higher percentage of marked constructions than unmarked,
which seems inconsistent with the claim that the unmarked precedes
the marked developmentally. Nor is there a clear increase in the
proportion of marked forms used by the advanced group as opposed to
the beginners, again suggesting a lack of a developmental sequence.
In addition, there does seem to be a developmental sequence in the
Inuit use of the unmarked constructions; that is, the advanced
students use the unmarked forms more than the other groups, surely
the opposite of what would be expected if the unmarked is the initial hypothesis. In fact, it looks as if these subjects are
gradually acquiring the more formal literary forms like (3a), rather
than starting out with them.[8] Thus, it appears that, at least in
certain cases, the presence of the marked construction in the positive data is sufficient for it to be learned, without the learner's
having first to acquire the unmarked version.[9]

   vii. Where L1 has the marked form and L2 the unmarked, the
usual prediction is that the unmarked will predominate in the initial interlanguage (see Table 1). I should like to propose the
opposite : not only might the marked case show up in the early
interlanguage but it may also be likely to persist, to remain as a
candidate for 'fossilization' (Selinker 1972). In situations like
these, the learner has had positive evidence in L1 for a marked
setting. He has, therefore, to notice the absence of some construction in L2, or to be able to view the L2 data totally divorced from
his L1 experience.[10]

   As a case in point, consider the opposite situation to that discussed by Mazurkewich (1984b), namely the English adult learning
French as a second language. From English, he will have had evidence
of the existence of marked forms like (3b). In French, however,
preposition-stranding is not possible.[11] Thus, forms like (4) do
not occur:

   4. *Qui as-tu donné le livre à?

I would suggest that learners who have already been made aware of
the possibility of marked constructions in L1 may also assume that
they occur in L2. Thus, the English adult learning French may
assume that sentences like (4) are grammatical in French.

Work by Tarallo and Myhill (1983) supports this prediction.
They show that preposition-stranding does cause interference, in
that ungrammatical relative clauses with stranding are judged to be
grammatical by native speakers of English learning various languages
where stranding is not possible. Muñoz-Liceras (1983) also found
that native speakers of English learning Spanish as a second language
at the beginner's level judged preposition-stranding to be possible
in Spanish, whilst students at the intermediate and advanced levels
did not. In addition, in an experiment involving grammaticality
judgments by native speakers of English learning French as a second
language (White 1983), ungrammatical French sentences with the
marked double-object construction (which is found in English but not
French) were judged to be grammatical both by the experimental group
and by a control group which consisted of native speakers of various
languages, who all also knew English. This suggests that the presence of a marked construction in L1 or in some other language that
one already knows is indeed a source of interference in L2.[12]

viii. The situation where both L1 and L2 have a marked parameter
or rule in operation is similar to the previous case. That is, if
evidence from both languages points to the need for marked forms, I
would suggest that the interlanguage will show the marked forms immediately, since otherwise the learner will have to ignore positive
evidence available in both languages.[13]

4. Markedness and parameters of core grammar; implications for
    L2 acquisition

So far, it has been argued that if 'marked' means a requirement
for positive evidence, as current proposals in linguistic theory
suggest, and if that evidence is available in L1, the L2 learner
will not necessarily be able to dissociate himself from his L1 experience when faced with another language which does not have the equivalent marked construction. In the case where the first language
has a more marked setting of a parameter of core grammar than the
second language, this could lead to a range of apparently unrelated
errors, which in fact stem from the inability to reset a particular
parameter.

In order to determine which settings of certain parameters of
core grammar constitute the unmarked case, learnability considerations are sometimes invoked (e.g. Chomsky and Lasnik 1977; Otsu
1981). Consider, for example, how the child determines the bounding
nodes for subjacency in his mother tongue. Subjacency is a principle of universal grammar which determines across how many nodes an

element can move. At most one bounding node can intervene between a
moved element and its trace, unless it has been able to move through
an 'escape hatch' like COMP. Thus, in the following sentences:

    5. a. Which race did Henry believe [$_S$'that Mary had won e]?
       b. *Which race did Henry believe [$_{NP}$the claim [$_S$'that Mary
           had won e]]?

(5a) is grammatical because the WH-phrase has crossed only one bounding node, in this case S', whereas (5b) is ungrammatical because both
S' and NP are bounding nodes.

Languages differ as to their choice of bounding nodes. For various reasons, Chomsky (1977) argues that the bounding nodes for
English are S', S and NP. The bounding nodes for Italian and French,
on the other hand, are NP and S' but not S (Rizzi 1982b; Sportiche
1982). [14]

We have here a case where a core principle may be set differently
for different languages; that is, there is parametric variation.
In certain cases, such variation simply implies differences in what
has be acquired in the languages concerned, whereas in other cases
the situation in one language may be more or less marked than
another. To establish whether markedness is a factor, one has to
consider whether the situation facing the child learning one
language is the same as that facing the child learning some other
language, with respect to the availability of the evidence.

On learnability grounds, we must assume that the situation is <u>not</u>
the same for all children trying to establish the bounding nodes for
subjacency. If only positive evidence is available to the learner,
then the situation for English is less marked than that in Italian
or French.[15] That is, if the child initially assumes that all
three of S', S and NP are bounding, then he will be right for English
and can work out that Italian and French do not have S as a bounding
node on the basis of acceptable sentences in these languages like
those in (6).

In other words, such sentences provide the child with evidence
which will force him to drop S as a bounding node.[16] If, on the
other hand, the child takes as the initial hypothesis that only one
or two nodes are bounding, then he will require negative evidence to
add bounding nodes. Supposing that the child learning English initially assumes that S is not a bounding node, he would be expected
to produce ungrammatical English equivalents of the sentence in (6)
and he would then have to be corrected in order for him to add S to
the bounding nodes. Thus, for a core principle like subjacency, the
unmarked case must be the assumption that all projections of S and N
are bounding, while cases which vary from this are more marked.

6a. Combien$_i$ [$_S$ as-tu vu [$_{NP}$ e$_i$ de personnes]]?

How many did you see (of) persons?

b. Ce théorème dont$_i$ [[une demonstration e$_i$ les a convaincu
     S NP
de l'utilite de ces techniques d'analyse]].....

This theorem of which a proof convinced them of the
utility of these techniques of analysis.........

c. C'est a mon cousin [wh$_i$ que [je sais [lequel$_j$ [offrir
                       S'   S   S'   S
e$_j$ e$_i$]]]]

It is to my cousin that I know which one to offer.

   Turning to second language acquisition, the Italian or French
adult learning E.S.L. has a marked setting for subjacency in his
mother tongue, in that not all potential bounding nodes are actual
bounding nodes. On most claims about L2 acquisition and markedness,
then, they will constitute examples of case (iii) in Table 1. That
is, if the L2 learner, regardless of his L1 experience, first adopts
the unmarked case, he ought to assume that the bounding nodes for
English are, indeed, S', S and NP. However, I would suggest that,
on the contrary, because S has been eliminated as a bounding node in
L1 in such cases, it may also not be treated as a bounding node in
L2. In other words, such situations should fall under case (vii) in
Table 2. This predicts a range of errors, all involving the assumption by the learner that S is not bounding in English. The interest
lies in the fact that such errors, apparently unrelated, could all
be explained as due to the failure to reset one parameter, from
marked to unmarked, that is, the failure to consider S as a bounding
node in English. Thus, French and Italian adults would be predicted
to find the English equivalents of the sentences discussed by
Sportiche (1982) and Rizzi (1982b) to be grammatical, even where
they are not.

   A similar area of interest, where a number of apparently unrelated phenomena fall out from the way a particular parameter is set,
is the so-called 'pro-drop' parameter. Languages such as Italian or
Spanish show a clustering of properties identified with this parameter (Chomsky 1981; Jaeggli 1982; Rizzi 1982a), including:

7. a. missing subjects

   b. free inversion in simple sentences

   c. occurrence of an empty category immediately
       following a complementizer (i.e. *[that e] violations

Examples of each of these are given in (8):[17]

8. a. e verra
        e will come

    b. e verra Gianni
        e will come Gianni

    c. Chi credi che e verra?
        Who do you think that e will come?

Non pro-drop languages like English do not show these properties, as
can be seen from the ungrammaticality of the English glosses· to
(8a), (8b) and (8c). It is assumed that the language learner faced
with a pro-drop language can decide that the pro-drop parameter is
operative on the basis of evidence from any one of the above areas
(Chomsky (1981). That is, evidence in the form of (8a), for example,
should lead the learner to assume pro-drop, so that (7a), (7b) and
(7c) are triggered together, even before the learner encounters specific evidence like (8b) and (8c).

The pro-drop parameter would seem to be marked, in that it requires specific evidence in the form of sentences like (8a), (8b) or
(8c). It presumably could not constitute the null hypothesis because, were the child to assume pro-drop in the absence of evidence,
he would require negative evidence to put him right.[18]. Thus,
Italian or Spanish speaking adults, for whom the pro-drop parameter
has been activated in L1, may have problems when faced with English,
if they fail to observe that English is not a pro-drop language.

Whilst the omission of subject pronouns has frequently been
observed amongst adult native speakers of Spanish learning English
as a second language (e.g. Cancino, Rosansky and Schumann 1974;
White 1977; Butterworth and Hatch 1978), other aspects of the parameter are only just beginning to receive attention in the L2 literature. In White (1985), the three aspects of the pro-drop parameter
discussed above are investigated. Native speakers of Spanish learning English as a second language do assume that it is possible to
omit pronouns in English, unlike French speaking controls.[19] In
addition, the Spanish speakers have problems in recognizing the
ungrammaticality of [that e] structures in English but they do not
assume that inversion is possible in simple sentences. Thus, whilst
they show evidence of difficulties in resetting this parameter, it
is not clear that all structures usually subsumed under 'pro-drop'
are equally affected.

5. Some developmental and pedagogical considerations

So far, the implications of marked parameters and rules in L1
for the early interlanguage of L2 learners have been discussed.

However, there are developmental implications here as well: if L1
has the unmarked form and L2 the marked (Table 2.iii), the learner
should eventually be able to acquire the marked forms, since the
evidence is available to him in L2. Mazurkewich's (1984a, 1984b)
findings that French subjects gained increasing mastery of the
marked double-object and preposition-stranding constructions is
evidence of this. However, if the situation is the other way round,
as in Table (2.vii), then the learner may have great difficulty in
ridding himself of the marked form in the interlanguage, unless he
is specificially corrected, and takes note of the correction. In
other words, I suspect that it might be harder for native speakers
of Spanish learning English to abandon pro-drop than it is for
native speakers of English learning Spanish to acquire it.
Similarly, it may be harder for native speakers of English learning
French to realize that sentences like (4) are not acceptable than it
is for native speakers of English learning French to learn that (3b)
is possible in English. In cases like those summarized in Table
(2.iii), the learner may have a better chance of ultimately achie-
ving native-like success than he does in cases like (2.vii), where
his interlanguage may remain like L1 in the relevant respects.
Thus, I assume that differences in markedness in the two languages
may affect the learning situation and the final outcome, though some
of my predictions are the opposite of previous claims in this area.

Cases like (2.vii) may point to a particular need for contrastive teaching or correction in the classroom in very specific circumstances. Much research in L1 acquisition has suggested that
negative evidence is non-occurring (Brown and Hanlon 1970) or in-
effective (McNeill 1966; Braine 1971) and the latter claim has also
been made for L2 acquisition (Dulay and Burt 1978). However, in
cases like (2.vii), the classroom may be the only source of the
necessary data to correct the learner. That is, if he has over-
generalized a marked construction from L1 to L2, there will be no
positive data in L2 to make him change his hypothesis. Thus, in
cases where L2 lacks a construction which is present in L1, it may
be helpful to point this out.

6. Conclusion

Corder (1979) argues that any theory of second language acquisition must account for the following:

   i.   the variability of the occurrence of interference phenomena

   ii. the similarity in sequences of development shown by
        learners of different mother tongues acquiring the same L2

   iii. the different magnitudes of the task of learning certain
        L2s in relation to certain L1s (especially unrelated
        languages)

I would like to suggest that viewing adult second language acquisition in the light of problems with resetting parameters, with
particular difficulties where L1 has a marked parameter and L2 does
not, goes a considerable way towards meeting all three of the above
needs.

Firstly, the variability of interference phenomena may not be
totally random but, rather, may depend on an interaction between
marked and unmarked parameters in L1 and L2. Whilst the suggestion
that interference plays a role in L2 acquisition looks superficially
like a return to the Contrastive Analysis Hypothesis, it is argued
here that interference is only to be expected in certain well-defined
circumstances. In addition, certain adults and a much higher proportion of children may have the ability to keep the data from the two
languages totally separate, thus avoiding the problems discussed
here.

Secondly, there will be many cases of overlap in the parameter
settings in L1 and L2, and also many cases where several different
mother tongues vary from the target language in the same way on the
unmarked/marked dimension, accounting for the frequently observed
similarities in the acquisition sequences of learners of different
language backgrounds (e.g. Dulay and Burt 1974; Bailey, Madden and
Krashen 1974; Cancino, Rosansky and Schumann 1978).

Finally, it may be the case that unrelated languages vary more
dramatically in terms of their parameter settings, or vary along
more parameters, leading to an increase in magnitude in the learning task in such cases. Thus, viewing adult L2 acquisition in the
context of a parameter-setting view of acquisition interacting with
a theory of markedness, may go some way towards explaining both the
similarities and the differences that have been observed in the
second language acquisition of adult learners of different mother
tongues learning the same target language.

FOOTNOTES

1. In fact, there is a potential contradiction in this definition
   of markedness. If the unmarked case is defined as the child's
   initial hypothesis, then he would be expected to adopt the
   unmarked case even where data from the language he is learning
   suggest that the marked case will be required. Thus, one has a
   developmental claim that the child acquires the unmarked as a
   necessary prior step to the acquisition of the marked. On the
   other hand, if the unmarked case is defined as the choice made
   by the child in the absence of evidence to the contrary, then
   one does not necessarily have a developmental claim at all.
   That is, given evidence to the contrary, the child might immediately assume the marked form, without first adopting the

unmarked. I will not decide between these two positions for
first language acquisition here. In general, the first definition is adopted by those working in the area of acquisition;
that is, markedness claims are assumed to translate into a developmental sequence. As we shall see, this is not necessarily
the case in second language acquisition.

2. However, see Mazurkewich (1984a,1984b) for detailed testing of
    the developmental claims of markedness in L2 acquisition.

3. As far as I am aware, no one has so far discussed within the
    markedness framework what factors will dictate the ultimate
    success or lack of success of adult learners.

4. In the current framework, it would presumably be subjacency
    (Chomsky 1973) or generalized subjacency (Baltin 1981) which
    would account for the facts of rightward movement in English.
    Nevertheless, Ritchie's findings remain valid; if Japanese does
    not have movement rules, then exposure to English provides the
    first occasion that subjacency could be activated for these
    subjects.

5. But see Stowell (1981) for arguments that preposition-stranding
    is not marked.

6. Notice how strong the markedness claim is in the case of the
    acquisition of preposition-stranding. Pied-piping (as in (3a))
    is uncommon in English. This means that if the language learner
    initially assumes the unmarked case, he has to ignore the many
    examples of forms like (3b) in favor of (3a), which presumably
    he could arrive at by the core rule of 'move ', already trig-
    gered by other aspects of the data. If learners do this, it
    would be good evidence for the strongest claims of markedness as
    having developmental implications (see note 1). In terms of
    their production data, children learning English as a first
    language do not seem to produce forms like (3a) before (3b).
    Krause and Goodluck (1982) attempt to establish the child's
    comprehension of such constructions; they find that three-
    year-olds have more difficulty understanding pied-piping than
    four-year-olds, the opposite of the developmental prediction.

7. However, Kellerman is largely concerned with idiomatic expres-
    sions and his definition of markedness is psychological rather
    than linguistic. Thus, it differs from the definition given
    here.
        Some other work by Mazurkewich (1984a) may be relevant to
    this issue. She investigates the acquisition of the double-
    object construction by the same French and Inuit subjects. They
    were asked for grammaticality judgments on sentences like:

i.   John gave the book to Fred
ii. John gave Fred the book

Cases like (i), where the verb is subcategorized for [NP PP] are
considered less marked than (ii), the double-object construction
(Williams 1981; Czepluch 1982). English allows both and, apparently, Inuktitut has neither, although it does have a double-
object construction, rather different from the English one.
Thus, the Inuit are in the position of having <u>no</u> relevant structure in L1 and learning an L2 where both unmarked and marked
forms are common. If markedness is equivalent to a developmental claim, they ought to prefer structures like (i) over (ii)
and they should show a developmental sequence in their acquisition of the marked forms. Mazurkewich did find that these
subjects showed consistently higher judgments of grammaticality
(95%) for forms like (i) over forms like (ii). However, the
latter were also judged to be grammatical in about 80% of the
cases and no clear progression was observable in the judgments
of the three groups (beginners, intermediate and advanced). It
is not clear that the [NP PP] forms were developmentally prior
from these results, though they do seem to be somewhat easier,
if there is a significant difference between the 95% response
level and the 80% level. Since the groups were chosen on the
basis of a Cloze test, and not on the basis of amount of
exposure to English, it could be that it was simply too late to
observe any possible developmental trends in this case.

8. 1 am grateful to Nancy Hildebrandt for drawing my attention to
these discrepancies.

9. See note 1.

10. The situation of having to notice the absence of some construction in L2 is similar to that of the L1 learner who overgenera-
lizes an optional transformational rule, as discussed by Baker
(1979). Given sentences like:

i.   a.   I gave the book to Fred
     b.   I gave Fred the book

and a transformational rule of dative movement, the learner
would be justified in assuming the grammaticality of (iib) on
the basis of (iia):

ii. a.   I reported the accident to the police
          *I reported the police the accident

There is, according to Baker, no positive evidence which could
rule out sentences like (iib) once the learner has produced them
(but see Mazurkewich and White (1984) for suggestions).

11. For certain exceptional cases, see Vinet (1979).

12. In the same experiment, the native speakers of English did not in fact, judge preposition-stranding to be grammatical in French. However, the control group did. This, in itself, is an interesting result as far as markedness is concerned, since, according to the claims summarized in Table 1, no learners should assume that the marked form is possible in L2 in such cases.

13. Sometimes language learners insert a pronominal copy into a relative clause, as in:

    i. The bed which the boy put the shoes under it is in the corner.

    It has been argued that language learners who do this when pronominal copying is not allowed in L1 or L2 offer an example of a case of the unmarked predominating over the marked in the interlanguage (Eckman 1977; Hyltenstam 1982). This would constitute a counter-example to my claims under (vii) and (viii). However, this is only a counter-example on the assumption that the presence of resumptive pronouns itself constitutes the unmarked case, which is not at all clear. In any case, the evidence from resumptive pronouns is somewhat confusing: at higher levels of Keenan and Comrie's (1977) 'noun-phrase accessibility hierarchy', pronominal copying is used only by learners whose L1 has pronominal copying but lower in the hierarchy (e.g. relative clauses on oblique NPs, genitives and objects of comparative) the insertion of pronominal copies occurs independently of what happens in L1 (Ioup and Kruse 1977; Gass 1979). Keenan and Comrie suggest that the insertion of pronominal copies is due to processing difficulties in such sentences, and this would seem to be true irrespective of the markedness question.

14. I leave aside the question of the bounding status of other nodes such as PP.

15. I assume that the burden of establishing marked options lies with positive evidence, since direct negative evidence is notoriously unreliable (McNeill 1966; Brown and Hanlon 1970; Braine 1971; Baker 1979). For arguments against 'indirect negative evidence' (Chomsky 1981), see Mazurkewich and White (1984).

16. The sentences in (6) correspond to Sportiche's (5c), (20) and (42). See Rizzi (1982b) for Italian sentences equivalent to (6c).

17. Taken from Rizzi (1982a). They are his examples (1), (2) and (3).

18. However, see Hyams (1983) for interesting arguments to the
effect that pro-drop constitutes the unmarked case and that
there is positive evidence available to the child learning
English which will eventually lead him to abandon it.

19. If 'pro-drop', in fact, constitutes the unmarked case, as argued
by Felix (1980) and Hyams (1983), then, on the views of markedness summarized in Table 1, the French controls would be in the
situation described in (1.iv); that is, their L1 and L2 would
both show the marked setting (i.e. absence of pro-drop) and they
ought to start out by postulating the unmarked case, thus
accepting pro-drop in English. Since they failed to do this,
instead recognizing the ungrammaticality of missing pronouns,
these results support the situation outlined in Table (2.viii);
in other words, having had evidence for the marked setting in
both L1 and L2, they immediately adopt this setting in their
interlanguage, suggesting that it is not the case that all L2
learners proceed through the unmarked structures as a necessary
step towards acquiring the marked.

REFERENCES

Bailey, N., C. Madden and S. Krashen. 1974. Is there a 'natural
sequence' in adult second language learning? Language Learning
24: 235-243
Baker, C. L. 1979. Syntactic theory and the projection problem.
Linguistic Inquiry 10: 533-581
Baltin, M. 1981. Strict bounding. In C. L. Baker and J. J.
McCarthy (eds), The Logical Problem of Language Acquisition.
Cambridge, Mass.: M. I. T. Press
Braine, M. 1971. On two types of models of the internalization of
grammars. In D. Slobin (ed), The Ontogenesis of Grammar. New
York: Academic Press.
Brown, R. and C. Hanlon. 1970. Derivational complexity and the
order of acquisition in child speech. In J. Hayes (ed), 1970.
Butterworth, G. and E. Hatch. 1978. A Spanish-speaking adolescent's
acquisition of English syntax. In E. Hatch (ed), 1978.
Cancino, H., E. Rosansky and J. Schumann. 1974. Testing hypotheses
about the copula and negative in three subjects. Working Papers
on Bilingualism 3:80-96. Toronto: Ontario Institute for
Studies in Education.
Cancino, H., E. Rosansky and J. Schumann. 1978. The acquisition of
English negative and interrogative by native Spanish speakers.
In E. Hatch (ed), 1978.
Chomsky, C. 1969. The Acquisition of Syntax in Children from 5 to
10. Cambridge, Mass: M. I. T. Press.

Chomsky, N. 1973. Conditions on transformations. In S. Anderson and P. Kiparsky (eds), A Festschrift for Morris Halle. New York: Holt, Rinehart and Winston.

Chomsky,. N. 1977. On Wh-movement. In P. W. Culicover, T. Wasow and A. Akmajian (eds), Formal Syntax New York: Academic Press.

Chomsky, N. 1980. On binding. Linguistic Inquiry 11:1-46.

Chomsky, N. 1981a. Lectures on Government and Binding. Dordrecht: Foris.

Chomsky, N. 1981b. Principles and parameters in syntactic theory. In N. Hornstein and D. Lightfoot (eds), Explanation in Linguistics: the Logical Problem of Language Acquisition. London: Longman.

Chomsky, N. 1982. Some Concepts and Consequences of the Theory of Government and Binding. Cambridge, Mass: M. I. T. Press.

Chomsky, N. and H. Lasnik. 1977. Filters and control. Linguistic Inquiry 8:425-504.

Clark, H. 1970. The primitive nature of the child's relational concepts. In J. R. Hayes (ed), 1970.

Corder, S. P. 1979. Language distance and the magnitude of the language learning task. Studies in Second Language Acquisition 2.1: 27-36.

Czepluch, H. 1982. Case theory and dative constructions. Linguistic Review 2: 1-38.

Dulay, H. and M. Burt 1974. Natural sequences in child second language acquisition. Language Learning 24:37-53.

Dulay, H. and M. Burt. 1978. Some remarks on creativity in language acquisition. In W. Ritchie (ed), 1978b.

Eckman, F. 1977. Markedness and the contrastive analysis hypothesis. Language Learning 27: 315-330.

Felix, S. 1980. Interference, interlanguage and related issues. In S. Felix (ed), Second Language Development: Trends and Issues. Tübingen: Gunter Narr Verlag.

Flynn, S. 1981. Effects of the reversal of principal branching direction (from L1 to L2) in L2 acquisition. Cornell Working Papers in Linguistics 2: 50-62.

Gass, S. 1979. Language transfer and universal grammatical relations. Language Learning 29: 327-244.

Hatch, E. (ed). 1978. Second Language Acquisition: A Book of Readings. Rowley, Mass: Newbury House.

Hayes, J. R. (ed). 1970. Cognition and the Development of Language. New York: Wiley.

Hornstein, N. and A. Weinberg. 1981. Case theory and preposition stranding. Linguistic Inquiry 12: 55-91.

Hyams, N. 1983. The pro-drop parameter in child grammars. In D. Flickinger (ed), Proceedings of W. C. C. F. L. II Stanford Linguistics Association, California.

Hyltenstam, K. (1982) Language typology, language universals, markedness and second language acquisition. Paper presented at the 2nd European-North American Workshop on L2 Acquisition Research. Gohrde, Germany; August 1982.

Ioup, G. and M. Kruse. 1977. Interference versus structural com-
    plexity as a predictor of second language relative clause
    acquisition. In H. Brown, C. Yurio and R. Crymes (eds), On
    TESOL 1977. Washington, D. C.
Jaeggli, O. 1982. Topics in Romance Syntax. Dordrecht: Foris.
Jansen, B., L. Lalleman and P. Muysken. 1981. The alternation hypo-
    thesis: acquisition of Dutch word order by Turkish and Moroccan
    foreign workers. Language Learning 31: 315-336.
Keenan, E. and B. Comrie. 1977. Noun phrase accessibility and
    universal grammar. Linguistic Inquiry 8: 63-99.
Kellerman, E. 1979. Transfer and non-transfer: where are we now?
    Studies in Second Language Acquisition 2.1: 37-57
Kiparsky, P. 1974. Remarks on analogical change. In J. M. Anderson
    and C. Jones (eds), Historical Linguistics. Amsterdam: North
    Holland.
Krause, M. and H. Goodluck. 1982, Children's interpretation of wh-
    constructions. In Y. Otsu et al. (eds), Studies in Generative
    Grammar and Language Acquisition. Tokyo: International
    Christian University.
Mazurkewich, I. 1984a. Dative questions and markedness. In F.
    Eckman, L. Bell and D. Nelson (eds), Universals of Second
    Language Acquisition. Rowley, Mass: Newbury House.
Mazurkewich, I. 1984b. The acquisition of the dative alternation
    by second language learners and linguistic theory. Language
    Learning 34. 91-109.
Mazurkewich, I. and L. White. 1984. The acquisition of the dative
    alternation: unlearning overgeneralizations. Cognition 16.261-
    283.
McNeill, D. 1966. Developmental psycholinguistics. In F. Smith
    and G. Miller (eds), The Genesis of Language: a Psycholinguistic
    Approach. Cambridge, Mass: M. I. T. Press.
Muñoz-Liceras, J. 1983. Markedness, Contrastive Analysis and the
    Acquisition of Spanish Syntax by English Speakers. Unpublished
    PhD. dissertation; University of Toronto.
Otsu, Y. 1981. Universal Grammar and Syntactic Development in
    Children: Towards a Theory of Syntactic Development. Un-
    published PhD. dissertation; M. I. T.
Phinney, M. 1981. Syntactic Constraints and the Acquisition of
    Embedded Sentential Complements. Unpublished PhD. dissertation:
    University of Massachusetts, Amherst.
Ritchie, W. 1978a. The right-roof constraint in an adult-acquired
    language. In W. Ritchie (ed), 1978b.
Ritchie, W. (ed). 1978b. Second Language Acquisition Research.
New York: Academic Press.
Rizzi, L. 1982a. Negation, Wh-movement and the null subject
    parameter. In L. Rizzi, 1982c.
Rizzi, L. 1982b. Violations of the Wh-island constraint and the
    subjacency condition. In L. Rizzi, 1982c.
Rizzi, L. (ed). 1982c. Issues in Italian Syntax. Dordrecht:
    Foris. Ross, J. 1967. Constraints on Variables in Syntax.
    Unpublished PhD. dissertation; M. I. T.

Rutherford, W. 1982. Markedness in second language acquisition. Language Learning 32: 85-108.

Schachter, J. 1974. An error in error analysis. Language Learning 24: 205-214

Selinker, L. 1972. Interlanguage. International Review of Applied Linguistics X.3.

Sportiche, D. 1982. Bounding nodes in French. Linguistic Review 1: 219-246

Stowell, T. 1981. Origins of Phrase Structure. Unpublished PhD. dissertation; M. I. T.

Tarallo, F. and J. Myhill. 1983. Interference and natural language processing in second language acquisition. Language Learning 33: 55-76.

Van Riemsdijk, H. A Case Study in Syntactic Markedness. Lisse: Peter de Ridder Press.

Vinet, M. T. 1979. Dialect variation and a restrictive theory of grammar: a study of intransitive propositions in a variety of French. Montreal Working Papers in Linguistics 13: 107-125.

White, L. 1977. Error analysis and error correction in adult learners of English as a second language. Working Papers on Bilingualism 13: 42-58. Ontario Institute for Studies in Education; Toronto.

White, L. 1983. Markedness and second language acquisition: explaining interference errors. Paper presented at the Canadian Linguistics Association annual meeting, Vancouver, June 1983.

White, L. 1985. The 'pro-drop' parameter in adult second language acquisition. Language Learning 35.

Williams, E. 1981. Argument structure and morphology. Linguistic Review 1: 81-114.

AUTHOR INDEX

329